COAL'S CONTRIBUTION TO UK SELF-SUFFICIENCY

Papers in this series are as follows:

Energy Paper No 8: International Gas Trade in Europe
Jonathan Stern

Energy Paper No 9: Coal's Contribution to UK Self-Sufficiency
Louis Turner

Energy Paper No 10: Gas's Contribution to UK Self-Sufficiency
Jonathan Stern

Energy Paper No 11: Electricity's Contribution to UK Self-Sufficiency
Richard Eden and Nigel Evans

Energy Paper No 12: Oil's Contribution to UK Self-Sufficiency
Robert Belgrave and Eileen Marshall

Energy Paper No 13: Conservation's Contribution to UK Self-Sufficiency
Mayer Hillman

Energy Paper No 14: The Economics of Energy Self-Sufficiency
Eileen Marshall and Colin Robinson

Energy Paper No 15: Natural Gas Trade in North America and Asia
Jonathan P. Stern

Energy Paper No 16: Energy Management: Can We Learn From Others?
George F. Ray

The opinions expressed in these publications are those of the authors and not necessarily those of the Institutions concerned.

The editor for the series is Margaret Cornell.

Louis Turner is a research fellow with the Joint Energy Policy Programme and with the Royal Institute of International Affairs. He is the author of six books and a number of monographs. His books include Oil Companies in the International System (3rd ed. 1983) and Middle East Industrialisation (joint with James Bedore, 1979). His last publication for JEP was the monograph, British Interests and the IEA (1982).

Copies in this series may be obtained from
Gower Publishing Company Ltd, Gower House, Croft Road, Aldershot, Hampshire GU11 3HR.

COAL'S CONTRIBUTION TO UK SELF-SUFFICIENCY

2nd Edition

Louis Turner

with a Foreword by Robert Belgrave

Gower

Published by
Gower Publishing Company Limited
Gower House, Croft Road, Aldershot
Hants GU11 3HR, England

Gower Publishing company
Old Post Road, Brookfield
Vermont 05036, USA

British Library Cataloguing in Publication Data

Turner, Louis
Coal's contribution to UK self-sufficiency.——
2nd ed.——(Energy paper; no. 9)
1. Coal trade——Great Britain
I. Title II. Series
333.8'22'0941 HD9551.5

ISBN 0-566-15016-1

Printed and bound in Great Britain by
Biddles Ltd, Guildford and King's Lynn

CONTENTS

TABLES AND FIGURES

FOREWORD

This report by Louis Turner of a Joint Energy Programme project was first published in the summer of 1984, when the coal strike was in its early weeks. It was based on information available in the autumn of 1983, largely from the Monopolies Commission inquiry into the coal industry and the evidence submitted to the Sizewell Inquiry by the Central Electricity Generating Board and the National Coal Board. The industrial action which led up to the strike was then beginning.

With the collapse of the coal strike we are now in a markedly different position and the way is again open for public consideration of what sort of coal industry we do want now. Indeed, it is essential that there should be such consideration to be followed swiftly by action, which will require not only the commitment of management and unions but also the support of Government and public opinion. Nothing could be more damaging than an attitude that the strike is over so let's all forget about coal. Since this report was first published, no more recent factual material that is relevant to the normal working of the industry has become available. It therefore seems useful to republish it with some small revision, but in particular with the addition of an epilogue which suggests, in the light of recent events, a course of action that government and NCB management could now take to make the coal industry viable.

The views presented in this paper are those of the author and I have joined him in the preparation of the epilogue. They are not necessarily the views of the institutions that sponsor or support the Joint Energy Programme.

We hope that this particular paper will make an opportune contribution to the development of policy by government and the coal industry, and to public debate on the future of coal.

Robert Belgrave
16 April 1985

SUMMARY

British policy-makers have hitherto paid little attention to the question of imports of coal because it was the competition from cheaper oil which was the predominant political issue. However, as the oil price rises of the 1970s have restored some of coal's former competitiveness with oil, it has become clear that British coal is now vulnerable also to competitively priced imports. Countries like the USA, South Africa and Australia have entered the international steam coal market in a major way and can deliver coal to NW Europe at prices below the cost of production of much British coal.

This study argues that the British industry is awkwardly placed, with a cost structure which is high by international standards, but is quite effectively protected by the high cost of transporting coal, especially to inland points of consumption. The core of the British industry's production should be competitive with imported coal under most forecasts. However, there is a significant tranche of British coal which could be increasingly vulnerable to imports if:

- operating costs rise significantly faster than international prices
- transport costs are reduced by some combination of a switch to larger coal carriers, the development of large-volume import terminals in the UK, or a gradual relocation of key coal-using plant toward the coast
- the value of sterling increases relative to the currencies of exporting countries.

The study points to the conclusions that:

i) there are positive reasons of a security and anti-monopolistic nature for keeping the import option open;

ii) imports of up to 30mt in the year 2000 could provide cheaper energy than corresponding domestic output;

iii) further study of the likely volume and price of internationally traded coal is required;

iv) total demand for coal in the UK is likely to remain in the 80-150mt range for the foreseeable future. At the lower level it may well be economic for the UK to be self-sufficient, but at the moment this is far from certain. The possibility of some imports should not be foreclosed by policy commitments until more is known about both the international and the UK cost structures.

COAL'S CONTRIBUTION TO UK SELF-SUFFICIENCY

The 1984-5 miners' strike was partly caused by the fact that British coal policy was in the melting pot. Mr Ian MacGregor's appointment as National Coal Board chairman in the summer of 1983 had been a visible sign of this. However, even before his appointment, the sheer size of the NCB's deficits meant that the issue of pit closures was inevitably coming to the fore, while the Monopolies and Mergers Commission report(1) was a pretty devastating condemnation of the way governments, the National Union of Mineworkers and the NCB management had led the industry into a financial quagmire. Even without the strike, a shift in attitudes would have been called for, whatever the political shade of party in power. What the abortive strike has done is to reduce the blocking power of the miners, should the NCB management and the government decide to impose radical changes on how the industry is run.

The strike tended to turn round the specific issue of pit closures. This study, however, concentrates on one rather neglected issue within the overall debate - that of liberalising coal imports. Over the years, British policy toward imports has been more or less protectionist. In February 1981, the Conservative Government actually reinforced this tendency by insisting that the Central Electricity Generating Board (the main consumer of coal in Britain) should stop coal imports at a time when the pit closure issue had become politically sensitive. Over and beyond this specific piece of protectionism, however, there was a tacit belief on the part of both the mineworkers and the NCB management that a surge of new investment would permit the development of coal mines which would be fully competitive with imported coal. If anything, the emphasis was on maintaining some kind of export potential, not on expanding import infrastructure. Implicitly, most British policy-makers accepted that the British steam-coal(2) market should be a closed one.

For many people, this seems a self-evident policy. After all, on some measurements(3), the United Kingdom has proven coal reserves of some 45 billion tonnes which would be enough to satisfy some 300 years of current consumption(4). How can imports be justified in such circumstances? Others have taken the opposite tack and have argued that the NCB's massive losses show that its production costs have been out of line with world price levels, and that there is thus a prima facie case for stepping up coal imports.

The coal strike has not changed this debate in any fundamental way. Coal imports certainly did not break the strike. That role was carried out by the recommissioning of oil-fired power stations and the continued production of coal by significant parts of the National Union of Mineworkers' membership. On the other hand, the strike has raised very serious doubts abut the wisdom of relying too heavily on indigenous coal supplies, so that the import option appears more attractive on security grounds. However, the strike took place against (indeed partly caused) a steady decline in the value of sterling and, though this has not justified Mr Arthur Scargill's claim that British deep-mined coal is the cheapest in the world(5), it very definitely has made imports an economically less attractive proposition.

The fact that developments during the strike seemed to pull in different directions is not particularly surprising, because calculating the precise future competitiveness of British coal is extremely difficult. In the first place, any such calculation rests on a number of fairly unquantifiable assumptions about how relative currency values may move. In addition, the underlying economics of the coal industry have been badly distorted by its recent history. This contrasts with the oil industry, where, although the existence of OPEC has created some distortions, there is a reasonably clear underlying cost structure which moves up steadily from the cheaply producable reserves in Saudi Arabia, through economically marginal fields in areas like the North Sea, to currently uneconomic deposits of heavy oil in the form of tar sands and shale oil. Only a complete collapse of OPEC's production controls would bring on stream a series of major oil provinces with substantially lower production costs than those of the marginal fields currently being developed within the non-OPEC world.

Coal is different. It was an unfashionable and decreasingly competitive fuel for so many decades (as cheap North American and Middle Eastern oil did their damage) that both exploration for and investment in new coal mines virtually ceased in many parts of the world. This has resulted in an ageing industry, in which it is still not uncommon to find operational mines which were actually opened in the last century. Over the last ten years, though, industry expectations have been revolutionised by the oil price rises of the 1970s. For the first time in many years, there was a general perception that coal could recapture markets from oil, and it thus made sense to start opening new mines which, because they had been discovered and developed using the technologies of the 1960s and 1970s, often proved considerably cheaper than the older mines they were replacing. So, unlike the oil industry where new fields have progressively become more expensive, the coal industry finds itself opening new mines which may actually produce cheaper coal than that from the bulk of existing mines.

The British coal industry is a classic example of this kind of dual development. On paper at least, new mines like Selby should produce coal cheaper than that from any other underground mine of the NCB. At the other end of the spectrum, there are the ageing pits of Wales which are massively uncompetitive by world standards. It is not argued here that the cold blast of competition should be allowed to sweep through the British industry regardless of social costs; coal is recognised as politically more sensitive than many other industries. However, what this study will argue is that serious thought needs to be given to the underlying cost structure of the British coal industry and to how that relates to the emerging picture of the likely structure of the international coal trade over the next thirty or forty years. British policy-makers have been guilty of very uncritical thinking about British coal policy in the past. The argument here is that we should not fall into the traps of the 1970s again. Will an accelerated programme of pit closures, combined with an increased investment programme in new mines, ever be enough to make the British coal industry fully competitive with imported coal? If not, what will the costs be to Britain of going for a policy of self-sufficiency in coal?

Recent history

The story of the twentieth century British coal industry is one of stagnation and increasing decline. Production peaked with the outbreak of the First World War, while employment in coal reached its own peak in the early 1920s at around 1.25 million(6). The industry more or less held its own through to the early 1950s, with a post-1945 production peak being reached in 1955 (Table 1). However, the onslaught of imported oil started biting deep into coal's markets in the 1950s, and coal was to lose nearly a quarter of Britain's primary inland energy consumption in the ten years 1955-65. By 1970, it had less than one-half of the British energy market, and, by the early 1980s, this share was barely above one-third.

The full story of the decline of the British coal industry is for others to chronicle(7). It is enough to note here that the decline in output went hand in hand with a massive decline in the industry's labour force, which by the early 1980s was one-sixth the size it had been at its zenith. On top of this development, there was a particularly important structural shift in how coal was used. In 1947, the prime uses of coal were, in descending order: domestic, industry, electricity generation, gas, coke ovens and railways. Electricity generation took just under 15 per cent of Britain's coal. Today, however, the picture is substantially different. The town-gas market has disappeared, primarily under competition from North Sea natural gas. Coal-fired steam locomotives only exist as tourist attractions. The domestic market for coal is now minimal, again resulting from the competition from natural gas. Industrial usage has also declined sharply, partly as a result of structural changes in the nature of British industry, but primarily as a result of coal's inability to compete with cheap fuel oil. This has left the electricity industry as the predominant market for British coal. Although there is some scope for coal to win back industrial markets, its future is heavily tied to the fortunes and strategies of the Central Electricity Generating Board (and its less important sister bodies like the South of Scotland Electricity Board). Coal has thus become a very narrowly based source of energy. Its over-reliance on the electricity sector leaves a distinct feeling of vulnerability.

As far as trade is concerned, the United Kingdom has historically been an exporting nation, with turn-of-the-century ports like Cardiff playing as proportionately important a role in world energy markets as Ras Tanura came to play in the early 1970s. In 1913, Britain was exporting 88 million tonnes of coal (which is thermally equivalent to just over 1 million barrels a day of oil - a massive trading achievement for those days). Exports were still significant during the 1930s, but the post-1945 peak was a mere 17.5 million tonnes in 1950. Since the early 1960s, exports have limped along at under 5 million tonnes per year, with a spurt to 9.4 million in 1981-2, but this was achieved at distress prices(8).

To all intents and purposes, coal imports did not become a serious political issue until the early 1970s. They only built up in the early 1950s as the British industry found itself unable to supply all of indigenous demand. From 1959 to 1969, coal imports were banned, but this was a minor part of the protection given the coal industry as it ran into the devastating competition from oil. Much more important was the tax put on heavy fuel oil, which effectively increased its cost by 25 per cent(9). However, by the early 1970s the purchase of cheap foreign coal started to become a real option for big coal users such as the CEGB (Table 2). Government response to this development has been to encourage and subsidise coal burning in power stations. The CEGB increasingly came under pressure to commit itself to taking given amounts of British coal (under the 1977 and 1979 agreements with the NCB) and, though it has managed to insist on financial reimbursement, it was formally requested in February 1981 to stop coal imports, which left it with the problem of disposing of Australian coal which it had already contracted for.

It is difficult to say how large coal imports might have grown to, if this mixture of direct and indirect trade barriers has not been erected. Certainly, in the early 1980s when sterling was still close to the $2 mark, it was clear that coal could be imported from countries such as Australia, South Africa and the United States at landed prices which could not always be met by fairly costed British coal. It was generally accepted that British coal had become uncompetitive with coal imported to power stations on the Thames. What was less clear, given the degree of subsidisation of the NCB, was how more widely competitive imported coal could have been if it had been competing against totally unsubsidised UK coal.

Past policies: how protectionist?

It is easy for economic liberals to be over harsh in their judgements of British coal policy since the mid-1950s. In fact, it is genuinely very hard to find comparable cases of major industries being so rapidly decimated as was the West European coal industry when it ran into competition from Middle Eastern oil imports after the Second World War. It would have been difficult for any politician to have left the British coal industry completely unprotected in the face of such competition. However, British attempts to slow the decline of indigenous coal should be put in the context of the times. In the case of the United States, it was not just the coal industry which was protected from the 1950s on by the quotas erected against imported oil, but the much more powerful US oil industry as well. Given the relatively greater vulnerability of the British coal industry, some protectionism was only to be expected.

What the shock of this competition from oil did do, though, was to blind the British coal industry to an important truth. Though it perceived imported oil as being its main source of competition, it tended to forget it still had to be ready to face competition from imported coal - and while attention was focused on oil, people generally failed to perceive that the competitive structure of the global coal industry was changing. In the 1950s, Britain competed most vigorously with the industries of continental Europe. By the mid-1970s, it was coal from sources such as Australia and the United States which was starting to matter.

However, the policy debate of the 1970s did not really reflect this changed competitive structure. Initiatives like the Plan for Coal(10) in 1974 or the 1978 Green Paper(11) were primarily concerned with allowing coal to regain markets as oil was priced ever higher by the OPEC cartel. The unpalatable truth, though, was that imported coal was in just as good a position to grab at least part of these expanded markets as the indigenous industry - a truth which was rammed home by the 25 per cent rise in the trade-weighted value of sterling between the winters of 1978/79 and 1980/81, which clearly made it attractive for coal users to consider expanding coal imports. These jumped from 2.3 to 7.3 million tonnes over this period. The political reaction to this increase occurred in February 1981 when the CEGB was forced to stop imports. This protectionism bought the industry time, but it could not disguise the fact that competition was coming not so much from the more heavily subsidised industries of France, Germany or Belgium, but from long-haul, unsubsidised sources such as the USA and Australia. This indicated that Britain's coal had become relatively high-cost by international standards and that, even given the protection afforded by high transport costs, a significant part of the British industry was now uncompetitive with imported coal. The fact that something had gone structurally wrong with the British industry was shown by the way that the NCB's deep-mining activities, which had become profitable (on NCB assumptions) in the mid-1970s, moved into the red in 1978-79 and, though oil prices again rose, continued to worsen, reaching a record £226m loss in 1981-2 (Table 3).

The current debate

Given this deterioration of the NCB's financial situation and the less sympathetic Conservative administration which took power in 1979, it is hardly surprising that a re-evaluation of British policy took place. One strand took the form of the investigation of the NCB by the Monopolies and Mergers Commission (henceforward the MMC). This was far from being an unprecedented step, since the MMC had carried out a similar exercise on the CEGB in 1981(12), and the thinking behind the relevant legislation (the Competition Act of 1980) always intended that it should do a series of similar investigations of other public bodies. However, the debate was given an additional fillip by the Sizewell B public inquiry, in which the CEGB sought to demonstrate an economic and technical case for constructing a pressurised light water nuclear power plant. Since the economic case turned quite heavily on assumptions about the future of the international coal industry, the relevant hearings sharpened the debate about where the British industry fits into the international context.

<u>The Monopolies and Mergers Commission Report</u>. The MMC investigation was the first authoritative critical evaluation to which the NCB had been subjected by outsiders. It is only partly relevant to this study, in that the MMC was not asked to comment on British coal policy in general, but was confined to an examination of the overall efficiency of the NCB. However, in carrying out this restricted task, the MMC team displayed a range of statistical detail not previously available to the public and also inevitably had to comment on various aspects of the NCB's trading policy and its competitiveness with imports.

The final report is a fairly devastating analysis of a corporation which had lost its way. The authors looked at the NCB's latest Development Plan which had been produced in July 1981 and concluded:

> On the information available to us, there is little possibility that the NCB will be able to operate without a deficit grant, let alone generate sufficient funds to finance any significant part of its own capital investment, before the end of this decade(13).

Although absolving the NCB's forecasters from the charge of having any worse record on energy demand than other recent forecasters, the report does suggest that management practices were somewhat behind the times, especially in the area of allocating investment costs to specific projects. In this particular case, comparing the financial performance of different mines or areas was confusing since interest charges were simply averaged over each tonne of coal produced throughout the NCB's operations. This meant that it was somewhat misleading for the NCB to present much of its financial performance in terms of 'operating' surpluses or, more often, losses. These 'operating' results were struck before taking interest (and social costs) into account, but as the industry becomes more capital-intensive this approach becomes increasingly confusing. Interest on the giant Selby project, for instance, should be over one-third of overall costs, but current NCB analysis will underestimate the costs of Selby production unless interest on the capital involved is included in the calculations(14).

This lack of financial sophistication was clearly behind what the report sees as the major strategic mistake of the NCB, namely, its perseverance with the investment goals of the 1974 Plan for Coal (which called for 40 million tons of new or modernised capacity), when demand projections were scaled down well below those underpinning the Plan, and when the linked pit-closure programme had fallen well below the Plan's commitments. The report suggests that the NCB became so fixated with the production targets of its development plans that it forgot about its financial obligations.

The end result of this strategic failure is an industry structure in which average proceeds in 1981-2 were only above the operating costs of between 40 and 50 million tonnes of NCB production (Figure 1). All the rest was unprofitable, even on the relatively generous NCB definition of 'operating costs'. In particular, what showed up is the 'high-cost tail' of NCB production, with operating costs starting to rise sharply once deep-mined production gets to around 80-90 million tonnes per annum.

So, what does all this mean for the debate about British coal trade policy? Certainly, there has been a growing acceptance that the landed price of imported coal effectively puts a ceiling on what the NCB can charge customers for equivalent kinds of coal. In the words of the MMC report(15):

> Since coal is a tradeable commodity and is in competition with other fuels, the NCB feels that the value of coal in the United Kingdom is affected directly or indirectly by the price of oil and more directly by the international price of coal.

This statement is supported by the assumptions built into the NCB's basic energy model, where it is specifically accepted that UK coal prices are linked to the price of imported coal, with the world price being given in US dollars at Antwerp, Rotterdam and Amsterdam (ARA)(16).

If imported coal is to set the price ceiling, there is still the question of how much of British coal is actually competitive with imports. In the words of the MMC report(17):

> The NCB told us that its aim was to meet as much of the potential demand in the United Kingdom for coal as it could meet economically. This, it explained, did not necessarily

imply that it should always plan to meet the whole of the potential United Kingdom demand for coal.

As we have seen, the MMC's terms of reference stopped it from making strong recommendations in this area, but the report clearly stated that some 10 per cent of the NCB's high-cost capacity needed closing on economic grounds(18), and it went further to speculate that, should the NCB genuinely set production targets based on economic viability, then they might 'be somewhat lower than forecasts of total United Kingdom demand' - that is, either higher imports or production subsidies would be needed. It quite correctly stated that such an issue is up to the government to decide. In particular, it floated the idea that the financial obligations laid on the NCB by section 1(4)(c) of the Coal Industry Nationalisation Act, 1946 may be too onerous, should it be decided for social or security reasons to continue limiting coal imports(19). Quite clearly, the MMC team saw the import issue as being of importance, even if strictly outside their specific remit. What they did reserve some sharp comments for was the NCB's coal export policy, under which coal has effectively been dumped on export markets. The use of the word 'dumping' may seem harsh, but dumping is deemed to take place when exporters sell goods abroad for less than they charge at home. In the words of the MMC(20):

> ... the broad position was that the NCB's revenue from coal sent to the Continental power stations was about £12.50 below the average price of £37.50 paid by the CEGB for NCB coal.

Although the MMC team listened to NCB justifications for this practice they concluded:

> We do not consider that over the longer term it (the NCB) should continue to rely on such low price sales for reducing surpluses of coal.

What the MMC report does do is to provide a pit-by-pit breakdown of operating costs for 1981-2, which finally allows outside observers to get to grips with the NCB's underlying supply curve (i.e. how much coal can be produced at what cost). Since these are NCB figures, they understate the true cost situation by some £2.60 per tonne, but this is not of vast importance. What one is faced with is a cost structure ranging from £21.60 per tonne at Bagworth in the South Midlands to £153 per tonne at Treforgan in South Wales. However, within these extremes, the important part of the picture is that the most efficient 36 million tonnes are produced at under £30 per tonne; the best 50 million for £32; the first 75 million for £38; and 100 million for £45 (Figure 1); in short, that only 40-50 million tonnes were being produced at a profit.

At this point the analytical problems really start. The CEGB gave evidence to the MMC team about the relative costs of imported as opposed to NCB coal (see Table 4). The 1982 picture was one of imports being well competitive at Thamesside power stations; being almost level-pegging at the Didcot station in Berkshire, which is some distance from a mine; and being generally uncompetitive in the Nottinghamshire coalfields, which are part of the belt in the Midlands where coal-fired power stations sit close to relatively efficient mines, with 'merry-go-round' trains linking the two ('merry-go-rounds' are unit trains specially designed for quick loading and unloading). However, in the Nottinghamshire case, even though imported coal bought on existing contracts was uncompetitive, spot purchases of coal could be bought sufficiently cheaply in the ARA region to make it competitive in the Midlands.

There are three main points about this import picture which need elaborating. First, the geography of British coal means that the relative competitiveness of imported coal varies very markedly across the country: transporting coal a hundred miles will add some 15 per cent to its cost. Secondly, the conclusions drawn from Table 4 will vary depending on whether one is talking about the current geography of the distribution of coal and electricity plant, or if one is looking at the next generation of coal-fired plant; in the latter case, one might conclude that the importing option would make greater economic sense than it does at the moment. Thirdly, the precise value of sterling matters a great deal in these calculations.

The point about geography matters since it rams home the fact that coal transportation becomes an increasingly expensive business once one is forced to transport it in relatively small units. Landing coal in the United Kingdom is one thing, but transporting it inland by rail starts raising its delivered cost quite substantially. So, though it was being landed by the Thames at prices which were cheaper than the production costs of at least 20 million tonnes of British coal, in practice much of this high-cost coal was safe because the cost of transporting the imported coal inland was too great for it to be competitive away from its most obvious landing sites. (The question of current import infrastructure will be returned to later.)

However, although existing geography seriously handicaps imports, it is clearly not unthinkable that future coal-fired electricity plants will be built on coastal sites. Whatever the difficulties imported coal may have in being competitive inland, Table 4 shows that coal was being landed at Thamesside power stations in the autumn of 1982 at prices virtually identical to those the NCB was charging inland stations close to NCB mines. (Between them, these stations consume around 6 million tonnes per year). Depending on where one thinks British coal prices are likely to go in comparison with international prices, a strategy of building future coal-fired power stations (or converting existing oil-fired ones) on coastal sites is clearly an option which needs economic evaluation.

The final observation to be made about Table 4 is that the calculations were done at a £1:$1.70 exchange rate, but since October 1982 sterling has declined precipitately. This has subtly changed the situation, with domestic coal having regained its position against imported US coal on Thamesside.

The trouble is that sterling is not the only currency which has been falling against the dollar. As it depreciated against the dollar by about one third between 1982 and early 1985, so did the South African Rand, while the Australian dollar fell by roughly one fifth. The result is that, while US coal was effectively being priced out of European markets, equally important international suppliers remained in roughly the same competitive position vis-a-vis the British industry as they possessed in 1982. Even at the sterling:dollar rates found in early 1985 (£1:$1.10), internationally traded coal was reaching North West Europe at prices (between $40-45 per tonne) which, when converted into sterling, were little changed from those of 1982. The major slide in sterling's value does not therefore seem to have significantly changed the position at the margins of the British industry. There is no particular reason why sterling should always move in step with the currencies of competing coal-producing economies and, given that the precipitate decline in sterling over 1982-1984 did not buy the British industry complete invulnerability from imports, there must be worries about what may happen should sterling strengthen significantly against the competition.

What is abundantly clear, though, is that the NCB's supply curve is such that the landed cost of imported coal will almost certainly remain an actual or potential competitive threat to part of the NCB's operations for the rest of this century. How serious a threat imports will be will vary with time, according to relative currency movements. In early 1985, a 20 per cent relative rise in the value of sterling would affect the competitiveness of some 20 million tonnes of British coal production. This sums up the problem for British policy-makers. They need to come to terms with an indigenous coal industry whose more expensive production is likely to be vulnerable to imports, the precise competitiveness of which will depend more on relative currency movements than on the underlying productivity of the coal industries concerned.

The Sizewell B inquiry. Given the increasingly dependent relationship of the NCB with the CEGB, the Sizewell B inquiry is of crucial importance to the British coal industry. Should Sizewell B be given an unequivocal clearance, then the CEGB would be more likely to switch to a high-nuclear strategy, and one could not then totally ignore scenarios for the year 2020 in which electricity's demand for coal is even assessed as falling below 10 million tonnes(21).

The CEGB case is that it will soon need to start constructing new generating capacity. Oil-fired plant is now extremely unattractive, except as a way of providing a high-cost, strike-breaking capability. The CEGB sees limits to the alternative conservationist route. As far as it is concerned, the choice is between nuclear and coal capacity, and, as far as its calculations go, there is really no contest. Under a variety of scenarios the nuclear option is a better economic bet than a three-unit coal-fired station (3 times 625 MW).

Now, it is not for the present writer to usurp the role of the Inquiry's Chairman, but some further basic points have emerged from the hearings. First, whichever way the eventual decision goes, the immediate contribution of coal to the British electricity scene is not in doubt, given that there is so much inertia in the system. Even if one assumes no stretching of nominal plant lives, there is enough coal-fired plant under construction to ensure that coal-fired capacity will continue to increase up to 1991, when coal will be providing 68 per cent of all available capacity (up from 63 per cent in 1983)(22). And the plant life of the capacity in place in 1991 will be such that the last plant then operating would not need decommissioning until 2026.

On the other hand, the end of the road for coal's role in electricity generation is starting to become a distinct possibility. Taking the high-nuclear variant of the CEGB scenario which most fits the assumptions behind the present BIJEPP project examining the UK's potential for energy self-sufficiency, the CEGB's coal burn could be expected to peak around 1995 (at 89 million tonnes(23). After that it would be all downhill for the coal industry, with the coal burn dropping by 22 per cent from 1995 to 2000, by a further 39 per cent by 2005, and then by 70 per cent over the next five years, leaving a coal burn of 12 million tonnes in 2010(24). Of course, these are the results of but one scenario, and a high-nuclear variant at that. On the other hand, it would only be a non-nuclear variant of our main scenario which would leave the coal industry feeling at all optimistic. If one assumes that we are in fact most likely to see a medium-nuclear future, then the coal burn still peaks in 1995, and falls by about a fifth in each of the next five-year periods, eventually stabilising at just over 50 million tonnes in 2010 (at which point a small but growing market for coal-fired combined-heat-and-power plants starts to emerge). We shall come back to these various scenarios at a later stage, but for the moment the implication would seem to be that electricity's dependence on coal is due to peak sometime over the next ten or fifteen years and that coal could become of totally marginal importance by the second quarter of the next century.

The Sizewell inquiry is not meant to analyse where whatever coal is needed by the electricity supply system should come from. In fact, however, a fair amount of evidence has emerged which throws light on this issue, and this has engendered a certain amount of heat in the cross-examinations of key witnesses.

The CEGB position is, in fact, remarkably bullish about the prospects of British coal against imports. In the words of Peter Hughes(25):

> The world background fuel analysis, coupled with the scenario projections of sterling exchange rate, show that the price of imported coal landed at the Thames estuary coal-fired stations will continue to increase in real terms. Eventually, in most scenarios, the price of imported coal delivered to the Thames estuary would rise to the point where its present (November 1982) 10 per cent price advantage over NCB coal, also delivered to the Thames estuary, would disappear and NCB coal would be competitive without subsidy.

There are two real strands to this argument: first, that international coal prices will escalate faster than those in the UK; second, that there will be a permanent depreciation in sterling which will reinforce the competitive position of the British industry.

As far as the CEGB is concerned, the picture regarding international coal prices is clear. Depending on the overall rate of world growth, the $70 per tonne steam coal price of 1980/1 should reach $88 in the year 2000 in a low-growth world, $105 in a medium-growth one, and $141 in a high-growth one. The annual price increases would average, respectively, 1.1 per cent, 2.0 per cent and 3.53 per cent(26). Such escalation is justified on the grounds that the international price for coal will be related to the costs and prices set by a handful of countries - specifically, the USA, Canada, Australia and South Africa. The CEGB argues that the cost structure in these marginal suppliers will have to rise over the long run if global demand is to be satisfied.

In the CEGB's words, these higher prices will come for the following reasons(27):

> The projected increases in the cost of labour, the capital costs per annual tonne of output from a new mine development, non-labour operating costs and the level of royalties and taxes in the three main OECD producing countries plus the Republic of South Africa, will combine to produce an increase in real terms in the price of coal at the mine. In addition, the necessary expansion of the transport infrastructure to allow increasing quantities of coal to be moved to the main centres of demand in Western Europe and the Far East will require substantial increases in transport rates.

These arguments have not gone without challenge, and the debate (to which we will return later) is highly relevant to any discussion of future British trade policy in coal.

The other particularly interesting argument by the CEGB is that the real sterling: dollar exchange rate will fall under four out of the five scenarios it considers (Table 5). In fact, because of the lag between its writing and the actual hearings, events have somewhat overtaken the projections, and in February 1985 sterling actually plumbed depths below those envisaged in even the most pessimistic CEGB scenario. Be that as it may, we only have to look back to 1980 when the rate was £1:$2.21 to realise we should not be swayed too much by immediate developments. After all, there is a strand of opinion today which believes that the dollar is 20 per cent overvalued, and it would be remarkable if there were not some reversal in the current weakening of sterling vis-a-vis the dollar. For the moment, though, we need only note that the exchange rate issue has come up in the Sizewell inquiry and that it is an integral part of the CEGB's incidental argument that imported coal is not going to pose too many problems for indigenous British coal.

In fact, the relationship between imported and NCB coal was not central to the Sizewell inquiry. It is the overall future of coal prices which matters, and the central CEGB argument is uncompromising about how insensitive its case is to a scenario in which coal prices stay static for the next ten years(28). Despite this, the CEGB case goes out of its way to show how the NCB should be able to expand. Even in the CEGB's middle-of-the road scenario (C) (see Appendix for outline of scenarios), it has NCB production staying stable at around 120 million tonnes per annum, rising to 150 million in 2030. In its high growth/high industrial scenario, it has NCB production climbing to 200 million tonnes per annum by that year (20 million from opencast). Imports are barely mentioned(29).

The eve-of-strike debate

The picture which emerges from the MMC study and the Sizewell inquiry is of an industry which is relatively high-cost when compared with the industries in countries like the USA, South Africa and Australia, but which is naturally protected by the high costs of transporting coal, particularly on land. However, events since the strengthening of sterling from 1979 show that the NCB cannot assume that transport costs will inevitably protect it from all import competition. After all, it is accepted on all sides that imports have been competitive with British coal on Thamesside (whatever the precise position today). On the demand side, the Sizewell inquiry is pointing to a long-term threat to the industry. There are now quite specific circumstances in which coal could start to lose ground heavily in the all-important electricity market sometime in the mid-1990s. Even in the most pessimistic scenarios, coal would not be wiped out overnight, but, by 2025, it could be of minimal importance in the market which is currently by far the most important.

However, if the picture of the British coal industry was being clarified, it still had sufficient murky areas to make the policy debate a lively one. For one thing, the MMC report helped in the creation of a supply curve of the NCB's activities (Figure 1), but it was still a snapshot of the situation in 1981-2 and gave no real indication of how that supply curve would move in the future. For instance, the MMC team accepted the assumptions made by the 1974 Tripartite Coal Industry Examination (government, NCB and unions) that closures of around 3-4 million tonnes per annum would be enough (if linked to new investments) to keep the British industry healthy(30). However, in retrospect, this is now a deeply suspect exercise, and such assumptions need justifying anew from scratch. Then, again, the optimists assumed that it would be possible to reinvigorate the British industry with a series of investments in mines such as the Selby complex which, by all accounts, should be delivering (floods permitting) at around £30 per tonne. A further capacity of between 10 and 13 million tonnes a year could reasonably be expected to have been developed by the year 2000 from new mine developments, such as those at Belvoir (2 million), South Warwickshire (2-4 million) and a further generation of smallish mines. In addition, a further 15 million tonnes per annum could probably be squeezed out of existing deep mines by judicious extensions. These figures are well below the NCB ambitions of the late 1970s(31) but pre-strike management was clearly studying what it thought its markets would actually be in the 1990s, and it appeared exceedingly unlikely that it would want to expand its current productive capacity.

What was not too clear was at what cost levels this post-Selby capacity could be provided. It seemed clear that Belvoir would produce coal at something above Selby levels, but South Warwickshire would fall somewhere between Selby and Belvoir. On balance, it appeared that this new capacity could be produced for costs nearer the £40 rather than £30 per tonne mark.

Given this initial picture of an industry which is marginally vulnerable to coal imports, precise future demand levels start to matter a great deal. Should demand contract more or less slowly toward the 75 million tonnes per year region, then there would be a prima facie case for arguing that the British industry could remain self-sufficient, since it would take extreme developments to render the NCB's core mines uncompetitive with imports. On the other hand, if demand for coal in the UK increased to the 150-200 million tonne level which seemed possible to observers in the mid-1970s, then the picture looked much more problematical. Such a level of demand would call for considerable new investment in mines whose economics had yet to be proved. Such demand might well call for the maintenance of some of the 'high-cost tail' of the British industry, but, as recent events had shown, this tail of some 10-20 million tonnes was already vulnerable to imports, though the vagaries of currency fluctuations might take some of the pressures off, or worsen them.

On the international front, there was still some uncertainty about how the coal price would move. This boiled down to a debate about which countries were likely to become marginal exporters of coal, in what time frame, and at what cost. Clearly, this question of the international price was important because it would help determine the precise level of import pressures on the British coal industry.

The strike

The 1984-5 strike did little to change the direction of the underlying debate about the future of the British coal industry. For most participants, the dispute was about the fate of the industry's high-cost tail. The importation of coal became an issue of short-run tactical importance, in that the NUM's position would have been strengthened if it could have cut off a key user like the British Steel Corporation from imported coal supplies. However, the strike was not primarily concerned with the import issue.

What the strike did do was make a mockery of any claim that being self-sufficient in coal somehow made the British economy more secure. There have been few international events with the impact this strike had on the UK, knocking, as it did, £4 billion off the balance of payments and increasing the Public Sector Borrowing Requirement by £2.75 billion. The 1973-4 oil shock probably came closest in its impact in recent times, and the response of the industrialising countries (Britain to the fore) was to take firm defensive action, which included the creation of the International Energy Agency, to ensure that the oil producers would not be able to wreak such economic havoc in the future. Given such action in the oil sector, the need to improve defences against further disruptions in the domestic coal sector was made more pressing by the intensity of the 1984-5 strike.

At the same time, the collapse of the strike has given the NCB management and the British Government the chance (or incentive?) to be rather more forceful in rethinking coal strategy than they might have been before the strike. Then, many people were aware that high-cost pits needed closing on economic grounds, but held back because of fears of the NUM's reaction. Now that the strike has taken place, such fears can largely be discounted, thus allowing decision-makers to worry more about the economic case for any given course of action than about the resultant political implications. This therefore should allow the debate to return to the exact nature of the British industry's supply structure.

Britain's supply structure

There is no doubt that the UK still has massive coal reserves (though Peter Hughes of the CEGB claims that 'proven accessible reserves' probably come to only 50 years of supply, rather than the 300 years which has become the conventional wisdom'(32)). Given the way that the NCB's exploration effort jumped in the late 1960s and (particularly) the early 1970s, these reserves should be increasing rather than diminishing(33). Reserves, however, are not at issue. What matters is the cost of producing them.

One of the ironies of the situation is that the cheapest coal available to the NCB is the most heavily constrained. This is opencast coal, of which 14.4 million tonnes was produced in 1981/2 (1.7 million of this being coking coal and anthracite), at an average operational cost of £27.51 per tonne(34). This coal is produced by a discrete part of the NCB, the Opencast Executive, and is heavily affected by environmental considerations. The target of 15 million tonnes per annum from opencast sites was set in the Plan for Coal, and has since become almost sacrosanct with the suggestion of the Flowers Commission on Energy and the Environment in 1981 that this target should not be exceeded in the foreseeable future. Given this constraint, the Executive has enough proved or partly-proved reserves to maintain output to the turn of the century(35). We are therefore faced with a situation in which the cheapest major source of coal available to the NCB is effectively constrained by environmental considerations. Perhaps as a result of these constraints, there is some doubt that opencast coal will remain relatively cheap, with the CEGB expecting opencast costs to escalate at 4 per cent per annum, which would imply that they would average some £55 per tonne by the year 2000(36).

However, this pessimistic analysis is countered in some circles with the argument that there are still plenty of potential sites for opencast operations, and that there is no evidence that the newer sites are becoming significantly less attractive in geological terms.

The key question which has to be answered is whether one can assume that the NCB's long-run history of cost escalation (1.5 per cent per annum over the last 25 years(37)) will suddenly improve. For instance, how realistic is it to assume that work practices can be permanently improved? However, even before the strike, the NCB was being formally required to reduce its average costs and its annual deficits, and a lot can obviously be done by closing down the 'high-cost tail' of grossly uneconomic pits. It can be argued by some industry observers that there is no a priori reason why costs should not be kept steady in real terms, but this clearly assumes that the NCB is free both to close uneconomic pits and to invest in major new profitable ones. A compromise policy of closures and penny-pinching investment will not be enough, because existing pits will become progressively more difficult to exploit and, other things being equal, real costs will continue to rise significantly.

The CEGB argues that the NCB should be able to limit cost increases to around 1 per cent per annum. Even so, the NCB still faces a situation in which its production costs will rise between 20 and 30 per cent in real terms by the year 2000. Applying these calculations to the existing supply curve, we are faced with the situation set out in Figure 2. This suggests that, should the NCB be supplying 100 million tonnes in 2000, it will be at a marginal operating cost of between £54 and 58 per tonne.

The only way this cost escalation can be avoided will be by a major productivity drive, which will probably have to go beyond a massive pit closure programme. Part of the solution will have to come from the opening of major new mines such as the ones at Selby (open, if not fully operational) and Belvoir (planning permission now granted). Selby is going to be a sizeable complex (10 million tonnes a year), but the authorised version of Belvoir is much smaller, and it is not clear where the next Selby-sized investment is going to come from. Yet to replace current high-cost pits a series of major new investments will be needed.

But will such projects revolutionise the NCB's prospects? The verdict has to be 'Not Proven'. If one is charitable and ignores the unexpected flooding of the first of Selby's pits (Wistow in August 1983), the economics do not look revolutionary. For one thing, the capital intensity of a coalfield like Selby means that its final costs cannot be very low. In the latest projections available to the MMC(38) operating costs were estimated to be £20.02, but interest charges were due to add another £12.68, bringing the final cost per tonne to £32.70, which should undoubtedly make Selby profitable, but will still be quite expensive by world standards. (It is still only some £10 cheaper than contractual prices for imported coal landed on the Continent in early 1985). In other words, it is cheap enough to be impervious to the challenge of imports, but is by no means guaranteed automatic export markets.

On balance, it would appear that the slow but steady rise in British coal costs will continue. Given the constraints on opencast mining, the UK does not have the option of substituting relatively cheap surface coal for more expensive deep mining. Any systematic improvement has to come underground, but there seem to be no innovations in sight which will revolutionise the production process there. Doubtless, the well-respected NCB underground engineers, like those in other industries, will be applying micro-electronics to their existing developments, but mines are considerably less flexible working environments than other kinds of productive enterprises on the surface. There is no question that the NCB's underground management is not technologically sophisticated, but the doubters point to the difficulties in improving productivity in the key area of transporting miners to and from the pit face at the beginning and end of shifts. However sophisticated control systems are, the physical dangers from speeding such underground transport mean that there is one key factor affecting productivity in which improvements will be exceptionally difficult to achieve.

In one area, that of subsistence, there are real grounds for suspecting that cost pressures on British coal will certainly be increasing. The cost of subsidence damage to the NCB rose fourfold in real terms to £90m. in 1983. The following year saw a drastic upsurge to £230m., about 70 per cent of which went to domestic property owners. There is no reason to think that popular resistance to subsidence is going to decline, and the development of new fields such as Selby and Belvoir is already constrained by the need to make a much greater effort to avoid causing subsidence in the future. If such standards were imposed on old fields, the impact on production from them would be quite severe.

What we are therefore talking about is a rapidly increasing social cost which is already coming to over £2 per tonne produced.

Given the pressures which are building on other forms of environmental disruption (sulphur emissions and so on), it is extremely hard to argue that subsidence costs will remain stable in real terms, let alone decrease. The overwhelming likelihood is that subsidence costs will increase over time, and that otherwise attractive NCB reserves will be sterilised because of public resistance (as has happened with the Belvoir project).

The picture is thus one of the NCB being stuck with a cost structure which is not amenable to any revolutionary developments. A determined blitz on its high-cost pits will undoubtedly bring the spread of costs between the high and low producers down to a more acceptable margin. However, the main strategy in controlling the upward drift will have to depend on how fast other fields like Selby can be brought into operation.

Taking all these considerations into account, it would appear reasonable to assume that the NCB's supply curve in the year 2000 will be much the same shape as today's. Obviously, there is going to be pretty heavy pruning of the high-cost parts of the existing structure; the future supply curve may therefore lack the sharp rise over the last 20 million tonnes of production which today's supply curve has. But there will have to be a massive new investment programme in coalfields with an economic structure not much worse than Selby's to avoid an inexorable rise in overall production costs.

International supply

Given this basic supply curve, the rest of this study is concerned with seeing how it relates to the likely development of international supplies. This is necessary in order to assess the costs and benefits of pursuing a policy of self-sufficiency. It uses as its base case a relatively conservative scenario which the Department of Energy labelled 'BL' in its evidence to the Sizewell inquiry (see the Appendix for details of this and other scenarios).

There is no real doubt that the international trade in steam coal is due to increase over the next couple of decades, but there is considerable debate about the precise magnitude of such an increase. There has been a general retreat from the over-enthusiasm of the WOCOL study(39) which had steam coal trade jumping from 98 million tonnes (53 million seaborne) in 1979 to 680 million in the year 2000. This kind of high projection now only shows up in extreme high-growth scenarios such as the CEGB's High Case(40), or in somewhat artificial exercises such as that of Ray Long for the International Energy Agency's Economic Assessment Service, which looked at the constraints holding back the speedy growth of this trade(41). Today, the range of estimates is considerably lower, with some now falling below the 300 million tonne range for the year 2000 (Table 6), though what is more important to note is the continued spread of the range of estimates. Shell, for instance, talks of the international steam coal trade growing between three and six times over 1981-2000(42).

The CEGB is offering a range of 266-603 million tonnes, depending on which of its scenarios one chooses. The NCB counters with a range of 250-509(43). By late 1984, there were even studies which implied levels of around 150 million only by the year 2000(44).

For the moment, however, the precise level of steam coal trade in the year 2000 is less important than getting an idea of the cost at which supplies will be coming on to the international market, because it is this which determines the pressures the British industry will come under to increase coal imports. The two most visible protagonists in the UK are the CEGB and the NCB which have had to expose their assumptions about the future of international markets to the Sizewell inquiry(45). In rather less detail, oil companies with international coal interests such as Shell and BP have also been forecasting future developments in international markets(46).

These three sources give rather different pictures of the future. At one extreme is BP which stresses the competitiveness of the international industry and the number of potential exporters waiting to enter the market, and concludes that prices 'of internationally traded coal in the long run should not increase by more than 1-2 per cent per annum in real terms and in a more realistic scenario would be more likely to remain constant in real terms'. This view is supported by Ronald Steenblik's evidence to Sizewell foreseeing landed prices in Europe as remaining between $38 and $63 per tonne. Similarly, Professor Richard Eden, who is co-author of one of the other studies in this BIJEPP series, is also willing to conceive of a scenario in which international coal prices stay stable at $60 per tonne(47). The CEGB is at the other end of the spectrum relying on quite detailed work on the likely future cost structure of each individual country. It makes predictions on this analysis which suggest that the price of marginal supplies of coal in international markets is going to increase quite sharply before the year 2000.

The CEGB argument is that international prices are likely to rise faster than internal British ones. Between 1980 and 2000, it expects the price of internationally traded coal, landed at ARA (Amsterdam, Rotterdam, Antwerp), to move from around $70 in 1980/1 to:

- low-growth case: $88 (1.1 per cent per annum)
- medium-growth case: $105 (2 per cent per annum)
- high-growth case: $141 (3.5 per cent per annum)(48).

There is something slightly counter-intuitive about this pessimism regarding the world cost structure; after all, are not large new surface mining regions still being brought into operation (Colombia; Powder River Basin in the USA)? Surely, this points to a healthier cost performance than that of the UK, which can only hope to constrain costs by opening Selby-sized underground mines.

The CEGB case stresses the limited number of countries

which will be doing the exporting (primarily, the USA, Australia and South Africa, with a few less important sources). As the CEGB case puts it(49), it:

> (expects) world traded coal prices to increase: (with) restoration of profit margins, progressive dependence on increasingly costly sources of deepmined coal for marginal and therefore price-setting supplies, opportunities for taxes, royalties and producer surplus and, in later years, demand pressure pushing up energy prices.

The NCB strongly opposes these CEGB arguments for a high-price international coal environment. It takes issue with both the CEGB and the Department of Energy (another source of scenarios assuming high-priced international coal). In dealing with the Department's estimates, it also stresses the counter-intuitive aspects of the CEGB's case which assumes:

> Rotterdam coal prices which in 5 out of 6 cases rise faster than oil prices to the year 2000 (and this is) difficult to believe given the limited availability of oil against the very large reserves of coal available(50).

The NCB suggests a range of international coal prices in 2000 of between $68 and $104 per tonne(51). Given that these contrasting forecasts indicate a very great degree of uncertainty, what sense can be made out of the competing arguments?

All observers agree that a limited number of exporters will be providing the bulk of internationally traded coal over the next couple of decades. There are arguments as to which one will be the most important, but the three suppliers which will be way ahead of any others will be South Africa, Australia and the USA. Countries like Canada, Colombia, Poland and, perhaps, the Soviet Union will be much smaller exporters. In the longer term other countries like Botswana, Indonesia, China and even Zimbabwe could enter the international market.

South Africa is a graphic example of how recent the latest interest in steam coal trade has been. This is not to say that coal has not always been absolutely crucial to the South African economy, but until a totally new port was commissioned at Richards Bay in 1976, steam coal exports had been minimal (just under 2 million tonnes in 1973). The other interesting point about the South African industry is that labour has been so cheap relative to capital that it was more economic for the coal companies to use Black labour underground than to invest in the plant needed for surface mining. This has given the mining houses good profits, and we are now faced with the paradoxical sight of the South African industry increasingly returning to the surface as Black labour becomes rapidly more expensive (they improved their wages by 15 per cent per annum during the 1970s)(52).

The quality of South African coal leaves something to be desired. Most of its reserves are high ash, and its higher grades also have a sulphur content of about 1 per cent (though that is not disastrously high). About half the reserves and most of its export coal are surface-mineable, even though the bulk of today's mining is underground(53).

The South African industry poses all sorts of interesting questions. For one thing, the local authorities are a bit like the oil producers, glad to have a valuable resource to export, but acutely aware of its impermanence - and South Africa's insecurity about exporting its coal runs deep since the country is so import-dependent on oil. Coal is its only real indigenous source of hydrocarbons. So, as various commissions have upgraded their estimates of the country's coal reserves, the authorities have increased the amount of coal which is allowed to be exported. At the same time, coal exports are now poised to become the country's second largest source of foreign-exchange earnings, so exports are going to be encouraged, even if the export ceiling is never totally dismantled. On the other hand, importers have to worry. For one thing, all exports go through one, super-efficient export port at Richards Bay, and though there seem to be no physical limitations on the further expansion of this port(54), there is the worry about over-dependence on this one terminal, should the country's political structure start to unravel at some future date. This is a particular problem for Western Europe, which is the most logical target for South African coal.

South African coal has also been becoming steadily more expensive, with overall costs rising faster than those in Australia and the United States over the last decade. However, they start from a low base and, though geological conditions should start deteriorating, there seem to be adequate coal resources available to sustain an expansion of exports at prices not too far removed from current levels. The picture therefore is of an industry capable of expanding exports (providing the demand is there) into the 80-120 million tonne range - though the higher figure would involve a certain amount of infrastructural expansion and would raise questions about whether high BTU low-sulphur coal might not be in short supply(55). On the other hand, South African coal is currently extremely cheap to produce and transport, with Shell reckoning that a new South African project could be landing coal in Europe at around $42 per tonne (12,000 BTU/lb low sulphur: 1982 $)(56). The CEGB, however, is much less sanguine about the future of South African export volumes and costs. It has worked closely with the Chase Econometrics' International Coal Trade Study(57) and in its central case, it sees South African exports being around 60 million tonnes per year in the year 2000, with costs of exports landed in Europe of between $94 and $132. This rather different scenario rests on the belief that the South African Government will continue to insist that only 5 per cent of available reserves will be allowed to be exploited for export; that a rapid increase of exports in the short run will lead to the South African industry moving up its supply curve more rapidly by the turn of the century; and that reserves of suitable low-sulphur high-BTU coal will become supply-constrained if the export trade expands too fast. Finally, the CEGB analysis rests on a strong conviction that South African mine-owners will be unable to keep the wage demands of their black labour force under control(58). However, labour costs only make up 14 per cent of underground and 6 per cent of new surface mining costs(59), so it is unlikely that labour inflation is going to price this coal out of world markets.

What we are faced with is two very different pictures of how the South African export trade is likely to evolve. The CEGB is quite right to stress that importers will probably set an upper limit to their dependence on such a politically complex supplier. It is also correct in pointing out the marginal role the South Africans assign to coal exports in relation to the key role assigned to coal in the home energy market. Inevitably the South African industry will move faster up its supply curve than if it were primarily an export industry and little else (as, say, the Colombian industry promises to be). For the moment, the CEGB analysis runs counter to some of the optimism expressed elsewhere about the potential of South African exports. The level of disagreement between the CEGB and its critics is of great significance. A world in which South Africa is a high-volume, low-priced exporter of coal will be much more difficult for the British coal industry, than one in which the CEGB forecasts come true.

Australia. There is also some controversy over the future of Australian supplies. At the moment, the country has a significant export industry (in both coking and steam coal - but this study concentrates on steam coal exports only). Just over half its production comes from surface mining (nearly 90 per cent in the case of Queensland). A new mine would be expected to produce coal landable in West Europe at around $60 per tonne(60). Reserves are quite healthy (see Table 7), but over half are in the sub-bituminous and lignite categories which are expensive to transport. In the more marketable anthracite and bituminous classifications, Australian reserves of 25.4 billion tonnes are only just over half those of the UK.

Though the CEGB and Chase estimates vary slightly, they both conclude that by the year 2000 in their 'best estimate' or 'Middle World' scenarios Australia's long-run marginal exporting costs will be the highest of the three main coal exporters (the CEGB put these as ranging between $87 and $100 per tonne when landed at Hamburg/Le Havre)(61). This relative pessimism about the Australian cost structure compares with studies like Long's for the IEA's Economic Assessment Service which has Australia somewhere between the marginal costs of South Africa (low) and the USA (high) in 2000(62). Even so, Long also puts the Australian marginal costs in the quite high $71-93 range, but it should be noted that his basic scenario is a relatively bullish one, assuming that there will be 645 million tonnes of steam coal on the oceans at the turn of the century - a figure which is now well at the top end of estimates. The NCB suggests that the marginal cif cost of Australian coal landed in North West Europe in 2000 is more likely to range between $58 and $77 depending on the precise scenario under consideration(63).

The CEGB/Chase case partly rests on a belief that Australia's reserves of high-grade steam coal are rather less bountiful than the optimists would argue. More significant for their case is the argument that heavy price increases are needed to give the companies and public authorities involved in the production and transportation chain an adequate return on their investment - for, without this, no adequate investment will take place. There is some support for this pessimistic analysis. For instance, the relative shallowness of the sea off Eastern Australia (the Barrier Reef) will put an upper limit on the size of coal carriers which can use Queensland ports, leaving the South Africans in a stronger position regarding shipping costs as the scale of bulk coal carriers goes over the 150,000-ton mark. There is some apparent additional support in recent attempts by the Australian authorities to set up a common bargaining front of the coal exporters to stop the coal importers (particularly Japan) from playing them off against each other. On the other hand, improving the Australian bargaining position is of little importance, unless they can form an informal exporters cartel with South Africa and the USA (an unlikely event). Certainly, a common Australian negotiating front will do nothing to improve the country's underlying cost structure, and once the exporters run into market resistance some part of the supply chain will have to drop its desired returns if Australian coal is to avoid being priced out of world markets. Already, in the current glutted market, we have seen the authorities dropping an export tax of one Australian dollar per tonne(64). Should Australia genuinely be the marginal supplier of coal to the international market in the year 2000, one would expect a wider response to market realities.

Accepting that there is some discrepancy in the available forecasts, it looks as though the range of possibilities is a maximum potential export capacity of between 80 and 120 million tonnes of steam coal in the year 2000. The price range is much more debatable.

The United States. There can be no doubt that the United States is currently the world's swing coal exporter. In 1981, after Poland and Australia had both taken themselves off world markets through their respective forms of industrial action, US exporters stepped into the breach, boosting their exports of coal to a record 99.8 million tonnes(65), or 86 million tonnes of seaborne exports, of which steam coal made 32 million tonnes(66). This relatively low figure reflects the fact that the US has traditionally been an exporter of coking coal, with steam coal exports only starting to climb significantly in the 1970s. By mid-1983, however, US exports were down some 38 per cent from an already slightly disappointing

1982. The chances are that final 1983 exports will be no more than 59 million tonnes (steam coal will be around 21 million of this)(67). This is against a world picture in which Poland and Australia are still expanding their exports, and South Africa and Canada are suffering much milder declines than the US industry, which has something like 100 million tonnes per annum mining capacity currently shut in(68).

The USA clearly has abundant coal reserves (Table 7), but it is the precise nature of these which is at the heart of the debate about where US coal prices will go. There are two considerations here: the geography and the sulphur content of these reserves. On the geography issue, Western Europe's supply of US coal has come from the US East coast, primarily from Appalachia. There are grounds for arguing that supplies of low-sulphur Appalachian coal will start to become constrained by the turn of the century, and at this point the picture gets complicated. The next most logical source of coal for Europe would be from the Illinois Basin, where it is available quite cheaply and could be transported by water down to the US Gulf and exported from there. The trouble is that this coal generally contains about 3 per cent of sulphur, which very much puts it into the high-sulphur category, and this raises all kinds of questions about the precise substitutability of low-priced high-sulphur coal as the price of low-sulphur coal edges its way upwards. Then, even further to the West, there are the massive, extremely cheap deposits in the Powder River Basin which straddles the Wyoming and Montana border. This is low-BTU and (generally) low-sulphur coal, with minemouth costs of around $15.50 per mtce(69). There is no way this coal will turn up in Western Europe, but its impact on mid-western electricity generation and, potentially, Pacific markets will have a knock-on effect elsewhere.

So, what do we make of all this? The first thing to note is that forecasts of exports from the United States are now starting to fall quite heavily. A cluster of recent forecasts put 1990 steam coal exports from the US at between 51 and 57 million tonnes, which means that steam coal exports will have to grow over 7 per cent per annum through the 1990s to get into the 100-120 million tonne range by the year 2000(70). The National Coal Association is making among the lowest forecasts for 1990, though this still remains higher than the NCB is willing to accept, even in its high growth case(71). However, on top of this disagreement about export levels, there is also another dispute about the marginal cost of US exports. As far as the NCB is concerned, these could vary between $49 per tce on its low trade scenario (involving some 26-32.4 million tonnes to North West Europe) and $94 (and a possibility of $106) in a high trade scenario involving some 55-60 million tonnes of steam coal to North West Europe(72). Working off a base case which is closer to the NCB's low, rather than high, case, the CEGB comes to much higher marginal cost figures of between $81.2 and $126 for, respectively, Appalachian surface and underground costs. Once again, the disparity matters because British coal will be under some import pressure under the NCB's estimates and none at all under the CEGB's.

A supply curve for the international coal trade?

In a pamphlet of this length, we must inevitably gloss over a number of important considerations. Clearly, there are forces working on international coal costs in two different directions. On the one hand, there are railroads, labour forces, and national governments which will seek to maximise their returns from the industry, but this is an important consideration only in those countries producing more cheaply than the marginal producer (actually, given the way coal importers will spread their purchases, there may well be two or three marginal producers at any one time). On the other hand, the industry is rather more competitive than may appear if one only thinks of the limited number of countries involved. For one thing, some of the exporters have political reasons for expanding sales (Poland, South Africa, even the Soviet Union). In these circumstances, prices will tend to get shaved to buy market share. Secondly, unlike oil, there is no State monopoly of coal production in any of the three main coal-exporting countries, and in general the international coal industry looks less oligopolistic than, say, the oil industry was in its prime.

Thirdly, the cheap reserves in the Powder River Basin and the possibility of new low-cost producers are reminders that the international coal industry has been far less well developed than the oil industry, and there is still time for new projects to come on stream which can have a significant impact on the late 1990s. There are other factors like the availability of ports which this study does not have space to analyse in any depth(73), though one might just note that one of the immediate responses to the trading boom of the early 1980s was a massive global expansion of port capacity, leaving a general position of over-capacity, though there are still some worries about the limited number of ports capable of taking really large-scale vessels. There is the further consideration as to whether international shipping costs will severely escalate over the next fifteen years or so. On balance, this appears unlikely. Admittedly, some recovery of freight rates will be needed for new ships to be ordered. However, it remains true that the coal industry is badly served by bulk carriers, and there is plenty of scope for improving transport efficiencies by moving into larger carriers. Given the way the shipbuilding industry has been migrating to newly industrialising countries like South Korea, it seems unlikely that construction costs will increase significantly.

But all this still leaves us some way from being able to construct some kind of supply curve for the international industry - and this is needed if we are to assess realistically the strength of future import pressures on the NCB. In Figure 3, we attempt a first approximation drawing primarily on the NCB and CEGB evidence to Sizewell but also on the work that Ray Long and Hugh

Lee have been doing for the Economic Assessment Service of IEA Coal Research(74). In some cases they have sufficiently broken down their estimates of the cif price of coal imported to North West Europe to give estimates of the cost structure of individual producing countries for given tranches of export volume. This is not necessarily a particularly elegant way of tackling the problem of the international supply curve, but it lays out the range of conventional wisdom available within the UK at the present moment.

This exercise does not produce a particularly coherent picture, for there are significant areas of uncertainty especially in the range of import costs which are of relevance to the UK industry. For the moment, it looks as though the international coal supply curve will have a significant upward slope in the year 2000, thus making it extremely important whether the international demand for traded coal lies in the 100 million or 400 million tonne range. However, putting it baldly, the CEGB, the NCB and the oil companies seem to be working off significantly different supply curves. Both the NCB and the CEGB accept that US reserves in the $90-120 price range will be increasingly needed as demand increases, but they are in substantial disagreement about the quantities of coal which could potentially be landed in the $60-80 range. Clearly, the major source of disagreement is over the potential of South Africa; here the two camps are in massive conflict, not just on price levels, but on the amount of coal which is actually going to be available for export. On balance, the CEGB scenario runs more against mainstream thinking than the NCB one. (This does not necessarily mean that the CEGB is wrong. It could just be that it is tempting to ignore the CEGB case, because so doing leaves quite a nice fitting curve from a first tranche of $50 American and South African coal, through intermediately priced Colombian and Australian reserves to apparently ample supplies of coal in the $100-110 range.)

We are left with three conflicting supply curves. There is the NCB one in which some 250 million tonnes of tradeable coal will be available to support a cif price in Western Europe of around $50 per tonne, before the price starts to rise toward the $90-110 region which will bring into play the rest of American and the more expensive Australian and East European coal. The CEGB scenario has a much more sharply rising initial supply curve, in which an international demand of around 200 million tonnes will push prices into the region of $100/tonne coal.

Finally, there is the 'oil industry' argument that the underlying competitiveness of the industry will bring new supplies of internationally traded coal into the equation at prices much nearer the $60 than the $100 range.

British supply in an international context

From this point, the present study's calculations are designed to illustrate the range of possibilities facing British decision-makers.

It tries to stick with scenarios which seem plausible, but the point is to explore variations around the more plausible cases to see how sensitive these central calculations are to the kinds of variation already obvious in the NCB-CEGB-oil company dispute about the international supply curve.

For the purposes of debate, the study analyses the variations round the case which the NCB was explicitly and implicitly putting forward at Sizewell. This assumed that the sterling-dollar exchange rate was as likely to rise as to fall, hence initial calculations were based on the £1:$1.50 rate which prevailed in the autumn of 1983 when this study was completed. It also argued that British coal costs should stay stable in real terms for the rest of this century. In so arguing, it disputed the CEGB estimates that British production costs will continue to rise in real terms in the 1-1.5 per cent per annum range to the turn of the century. After all, it was argued, NCB management has now been given clear government instructions to bring down the industry's average costs. It can see where new coalfields can be opened at costs toward the lower end of the current supply curve. It also sees no reasons why opencast costs should escalate drastically in the mid-term future. Finally, it was asserted that it should be quite easy to produce an extra 15 million tonnes per year from existing pits at costs which are also at the lower end of the NCB supply curve.

It does not matter precisely how plausible the reader finds this picture, though one would certainly expect the Thatcher-backed MacGregor management team to be rather more successful in controlling costs than the NCB administrations of the 1970s, who were managing in a more carefree era. What does matter is that it gives a comprehensible scenario to analyse, which in some areas (such as its projections of international coal prices) is not unduly favourable to the British coal industry.

Taking the estimates made in Figure 3 of the landed cost of international coal in North West Europe, we then throw into the equation the costs of further handling and transporting this coal to the UK and, on occasion, inland as far as the Midlands coal belt. These calculations have been worked backward to give the price below which successful tranches of British coal must keep in order to stay competitive with imports(75).

Such calculations indicate that the British industry has a sizeable core which should, under most scenarios, be impervious to international competition. Even under an extremely pessimistic scenario combining a strong pound (£1:$1.75) with a failure to control British costs (a 2 per cent per annum increase in real prices), there should be some 40-50 million tonnes which could still hold their own against imports costing as little as $60 per tonne on the Continent (which is the most generally accepted minimum price at which substantial quantities of coal could be landed in North West Europe in the year 2000). Under most scenarios, however, the industry should be much more widely competitive than that, and the interesting question is not whether the British coal industry as a whole can compete with imports, but how the higher cost ranges of British coal will fare under various alternative assumptions.

Even assuming no real price rises for British coal and a £1:$1.50 exchange rate, the British industry is clearly sensitive to its ability to control costs; the movement of sterling against other currencies (particularly the dollar); the precise shape of the international supply curve; and the overall size of the traded steam coal market in the year 2000. In addition, it is somewhat vulnerable to a reduction in transport costs which could come from the opening of transshipment facilities on the British coast capable of taking the giant coal carriers which currently can land only on the Continent. The combination of these factors leaves a range of uncertainties which will make planning for the British coal industry relatively difficult.

International supply. In practice, it looks as though the most important uncertainty is about precisely how much coal there is in the international system which could be brought to Europe in the year 2000 at various delivered prices in the range between $60 and $100 per tonne(76). Assuming that current projections of international demand in the year 2000 are increasingly falling toward the 300 million tonnes region (Table 6), it is striking that it is precisely the 200-300 million tonnes area in which the NCB and the CEGB are projecting international costs to rise particularly steeply toward the region of $100 per tonne for coal landed cif in North West Europe. Under various assumptions about the future strength of the dollar, it looks as though there could be about 250 million tonnes capacity in the year 2000 which could enter trade at about $60 per tonne. This capacity would be found in the US East coast (around 50 million tonnes), Australia (50 million) and, the NCB argues, South Africa (110 million), with Colombia contributing another 50 million somewhere above this range(77). Even if one rejects the CEGB/Chase case about the constraints on cheap South African coal, current projections still see a rapid cost escalation in the next tranches of Australian and American coal which appear to come into the equation at between $90 and $100 a tonne - a price level which would not cause too many problems for the British coal industry.

There is something intellectually unsettling about the apparent absence of any substantial tranches of coal with marginal costs capable of bringing them into the international market at prices between $70 and $90. It could be that the geography of the world coal industry is peculiar and that there will indeed be a steep jump in marginal costs once capacity in the $60 region is exhausted. This would mean that the international price would rapidly rise to the $90-100 region where it is generally accepted that plentiful reserves exist. If this is so, then the British industry will generally be in a reasonably secure position in that, under most demand scenarios, the world price will be nearer $90 than $60. However, the year 2000 is still some 15 years away which leaves plenty of time for new exporters (Botswana? Indonesia? China? the USSR?) to establish themselves in this $70-90 range, thus giving the British industry rather more trouble.

This is not an unimportant debate because the sensitivity of the British industry to unfavourable developments is likely to be drastically different in the $60 and $90 regions. To take but one example, consider the sensitivity of the British industry to its relative success in controlling costs. If one assumes a $60 price in 2000, then the competitiveness of some 40 million tonnes of British coal would be affected by whether British costs stay stable in real terms or increase by 2 per cent per annum. At the £1:$1.50 exchange rate some 125 million tonnes would be competitive with imports on the no-inflation assumption. On the inflationary assumption, no more than 80-90 million tonnes would be competitive. However, if the international price were in the $90 region, then the British industry would remain relatively insensitive to inflationary pressures, with only a marginal decrease in competitive production under the high inflation assumption of around 10 million tonnes (leaving some 125 million tonnes capable of fighting off imports).

Clearly, policy-makers face considerably different problems in the two cases. A variation of 10 million tonnes could well be handled by advancing pit closures already in the planning pipeline, but a potential swing of 40 million tonnes would pose much more serious problems and would need handling in a totally different manner. The logic of this is that it is absolutely essential that British policy-makers clarify their thinking about the nature of the international supply situation. No coherent plans can be made so long as it is possible to have projections varying as wildly as do those of the NCB and the CEGB in the case of South Africa. In particular, it is important to get a grasp of whether there are potential new entrants which could be providing an extra 50-100 million tonnes of coal by 2000 at prices closer to the $60 than the $100 range. If these did emerge, then the sensitivity of the British industry to import penetration would be quite markedly increased.

Escalation of British production costs. Of the other factors potentially affecting the British industry, the two main ones are the impact of currency fluctuations and the relative cost escalation. The latter phenomenon is potentially under national control, while exchange-rate changes will hit the British economy more randomly.

As we have said earlier, the NCB failed during the 1970s to generate sufficient productivity improvements to offset rises in its wage bill. The result was a steady rise in real costs of 4 per cent per annum. On balance, one would expect this record to improve for the rest of the century. The era of Plan for Coal, with its emphasis on expansion, is over and there is now much more emphasis on controlling costs in order to improve the industry's general competitiveness. It now looks as though the 'high-cost tail' of grossly uneconomic mines will be phased out quite rapidly, and the opening of relatively low-cost mines such as Selby should also contribute to lowering (or at least controlling) average costs. Assuming currency stability therefore, inflation is not going to jeopardise the existence of the entire British industry, though it could be enough to reduce the viable industry below the 100 million tonnes level.

Currency fluctuations. The expected range of fluctuations in the sterling-dollar exchange rate should have an impact of much the same magnitude as variations in the relative cost escalation of the British industry. Assuming that the rate should fall somewhere between £1:$1 and £1:$2 in the year 2000, then we are once again faced with an impact on the most marginal 30-40 million tonnes at the $60 price, and 10 million at the $90 price. At the lower price range (and assuming no real price rises) one would expect around 130 million tonnes to be competitive at the £1:$1.50 exchange rate, against 140 million tonnes should the pound fall toward the £1:$1 level, and about 100 million tonnes should the pound move in the opposite direction toward the £1:$2 region. In addition to giving an idea of the range of the possible variation, these figures show that currency movements have an asymmetrical effect, with a strengthening of sterling putting competitive pressures on a larger tranche of British coal than would be relieved by its depreciation. On balance, the odds are that the long-term depreciation of sterling against the dollar will continue as the effects of North Sea oil wear off, but this is by no means guaranteed and if a renewed strengthening is still plausible, then the UK should work on the assumption that competitive pressures on the British coal industry could increase quite significantly.

The transport barrier. Much of this relatively optimistic picture of the ability of the British industry to compete with imports rests on the effective protection afforded it by the high costs of transporting coal. If these costs should fall for any reason, then British coal production will come under stronger pressure from imports. The extent to which such a cost reduction can occur breaks down into three separate areas of concern. First, there are questions about the future costs of long-haul sea transport needed to bring American, South African and Australian coal the unavoidable thousands of miles to North West Europe. Second, there are the handling and on-shipment costs involved in moving coal from the Continental super-ports like Rotterdam or Le Havre to the British coast. Third, there are the transportation costs involved in moving coal from the coast to internal destinations. As the CEGB claims in Table 4, moving coal from the Thames to the Midlands is almost as expensive as getting it from the Continent to the Thames.

The future of international coal shipping costs is open to some debate, with the CEGB being particularly bearish in arguing that they will rise substantially by the end of the century(78). There is something in the argument that the next generation of large bulk carriers is not going to be built unless returns improve considerably from their currently depressed levels. On the other hand, coal carriers are markedly smaller than tankers employed in shipping crude oil, so one must assume that there will be pressures on coal shippers to move toward larger vessels to improve their scale economies. In 1982 the dollar-per-tonne transatlantic charter rate could vary by as much as three to one, depending on whether the coal was transported in a 26,000 tonne capacity carrier ($17.89) or one with a 92,000 capacity ($5.63)(79). 150,000 dwt 'super-colliers' could easily be built using today's technology, and these should cut another 20-40 per cent off current ocean transport rates(80). As they do move toward larger carriers the pressures of long-distance shipping costs will be reduced, but the enthusiasm with which shipping companies will make the necessary investments will depend on the underlying buoyancy of international coal markets. The paradoxical effect of all this is that unit shipping costs could well be highest in the low-trade scenarios (less than 300 million tonnes) because shippers will not have the incentive to invest in a new generation of super-colliers. The full impact of improved scale economies is most likely to be felt in the high-demand scenarios as one more factor limiting the ultimate upward movement of prices. However, it is precisely in those scenarios involving well over 300 million tonnes that the international price is likely to be high enough to avoid causing the UK industry too much trouble, so the downward pressures on shipping costs will be relatively irrelevant.

Moving to the other end of the transport chain, it is unlikely that much will happen over the next twenty years to revolutionise inland transportation costs. As long as rail is the chief mode of inland coal transport, no cost revolutions can be expected. The current merry-go-round train principle would be used between ports and inland destinations, but it is unlikely that the costs of such unit trains can be significantly reduced in real terms. Nor can there be much hope of lower costs from new technologies such as coal slurry pipelines. Although there is considerable debate about the relative merits of coal:water, coal:oil and coal:methanol pipelines, precious few have actually been built, and they suffer the ultimate disadvantage of inflexibility(81) which, in the British environment, must probably rule them out. After all, one of the conclusions of the present study is that it is necessary to plan for some unpredictable fluctuations in import competitiveness. It would therefore be unwise to tie British fortunes too closely to a mode of transportation which is totally inflexible. Unlike merry-go-round trains which run regardless of whether they are being loaded from an inland mine or from a coal-carrying ship, a pipeline is left as a white elephant if economic circumstances change and undermine the logic on which it was originally constructed.

This leaves a reduction in the on-costs from the Continent to Britain as the most likely way in which the competition from imports is likely to be increased. At the moment Britain's port facilities are enough to import around 12 million tonnes per annum(82), but most of this would have to be in vessels far smaller than the current optimum sizes which ports like Rotterdam can accommodate. Britain does possess a few ports - Hunterston in Scotland, Port Talbot in Wales and Redcar in England - which are capable of handling vessels of the desired size of 100,000 tonnes and upwards (Table 8). Immingham, which the NCB uses for coal exports, is the next most well endowed port capable of handling vessels in the 60,000 tonne region.

The trouble with such ports is that they tend to be in the wrong location (Port Talbot, which is too far from most English power stations) or already under the control of the British Steel Corporation (Redcar) or the NCB (much of Immingham). In the case of the NCB at Immingham, we are dealing with facilities specifically designed only for exporting. In an economically rational world, there might well be a case for removing these facilities from the NCB's control, given that Immingham is one of the most logical points at which coal imports would enter if they were to be competitive with the Midlands coalfields. An independent dock operator would then be free to decide at what point to build import as well as export facilities.

The remaining British ports are only capable of handling much smaller vessels. Even in the case of the Thamesside power stations, the maximum size of vessel which can be handled by their discharge facilities is only being raised from 11,500 dwt to 18,000 dwt - still a sub-optimal size. The reason why it is most economical to land coal first in ports like Rotterdam is that it makes sense to use bulk carriers for the long haul, breaking the cargo for transporting in smaller ships to the UK. The way to bring this cost down would be to construct terminals in the UK capable of handling bulk carriers in their own right, so that the whole transshipment exercise is cut out. The CEGB claims(83) that it has studied all the deep-water estuary sites and has developed a portfolio of sites which could be developed quite rapidly (2-3 years?) if need be.

The development of one or two such import terminals would have an incremental impact on the British coal industry, which would initially maintain some special protection from the fact that the geographical location of existing coal-fired plants is inevitably skewed toward existing British mines. This means inland locations, which would still leave quite heavy on-costs for the transportation of imported coal from the coastal terminals. The protection which would initially be lost would be the extra shipping costs from mainland Europe to the UK, which would need to be balanced against the cost of constructing import terminals of the necessary capacity. The extra costs involved in such an investment would very much depend on the extent to which the importing capacity could be located within an existing port, which would thus minimise the extent to which major channel deepening etc. would be needed. On land, the incremental costs would be minimal since the rail capacity needed would be identical to what is already used between inland mines and their customers.

Thus, depending on the precise cost of developing the necessary terminals, a policy of taking bulk carriers directly into the UK would save something approaching £7 per tonne, which in this study's base case (£1:$1.50 and no real inflation) would undercut a marginal 15-20 million tonnes if landed prices established themselves in the $60 per tonne region. In the $90 region, the impact would be minimal.

Should the necessary investment in bulk importing capacity be made, there would follow the almost unquantifiable question of how the existence of such capacity would affect future locational decisions of the significant coal users. For instance, if the Sizewell 'B' decision steers the CEGB away from a heavy nuclear stance for the 1990s, then the Board will have to decide where to locate the coal-fired plants which will be needed to replace plant which will have to be closed during that decade. The CEGB might well be tempted by the strategy of relocating some of this coal-fired capacity near the main import terminal(s). This could be a risky strategy should the import terminals be located in one of the more peripheral parts of the British coal market (Wales or Scotland?). Should the forecasts of import prices go awry, then the new capacity would be left stranded at an expensive distance from the cheapest sources of indigenous coal. It would therefore probably make more sense to develop new stations alongside a major import terminal on the English East coast, where the distance from the major Midland coalfields is relatively short, thus providing a reasonable fallback position should international prices prove higher than expected(84).

There is nothing particularly unusual about the idea that coal users could start to relocate plant close to major import terminals. This is, after all, a development which has already taken place in industries like steel (for its own particular sectoral logic). What it would mean, though, is that the British coal industry would come to lose virtually all the extra protection it has had from internationally traded coal in North West Europe; and once a plant has been relocated away from British mines to the coast, transport economics start to weigh against the British industry in that, not only does a coastal location reduce the landed cost of foreign coal, but the costs of transporting coal within the UK mean that the indigenous industry becomes even more expensive.

Accepting that coal users are unlikely to turn their backs on British coal so completely that they would relocate all their coal-using plant at the first opportunity, such relocation would put the British industry under considerably greater strain. Taking the £1:$1.50/stable British cost scenario once again, and assuming that the international price is in the $60 per tonne region, then it could be that a further 30 million tonnes of British coal would become uncompetitive (over and above the 15-20 million potentially rendered uneconomic by the initial development of bulk importing facilities). The industrial logic following on this latter development could leave only about 75 million tonnes of British coal strictly competitive with imports. However, once again, this is all highly sensitive to the final level of international coal prices. At $90 per tonne, the long-term effect of developing the bulk import of coal to the UK would be minimal.

The demand equation. The two remaining major unknown variables needed for the British coal equation are the likely demands for, respectively, internationally traded and British coal. On balance, the British figure is not particularly crucial, though obviously the NCB needs to get its estimates right to avoid maintaining persistent productive overcapacity. What is much more important is to get a sense of what the international demand is likely to be, and then to assess how this demand is likely to interact with whatever is the accepted international supply curve. The problem, though, is that there is still quite some disagreement on the nature of that curve. But, however inconclusive such an analysis may be, it remains absolutely crucial, because a British industry faced with $60 per tonne coal will be in a completely different position from that if it were faced with $90 per tonne coal.

For the moment, it seems unwise to become too wedded to any single forecast of the international market for coal over the next twenty or thirty years. It now appears likely that forecasts made in the heady days of 1980/1 are proving to have been over-optimistic. In the short run, coal demand has been very sensitive to the recent slowdown in the world economy, with spot prices for coal landed in North West Europe falling as low as $30 per tonne on occasion during 1983. What are more worrying are signs of a general downward reassessment of coal's prospects in key sectors. Plans for prototype coal liquefaction and gasification plants have been scrapped throughout the world. Optimism about coal's penetration into industrial markets has fallen in parallel with the decline of oil prices, which, though not restoring a strictly comparable price to coal, have still fallen enough to make some conversions of oil-fired boilers uneconomic. The Chase Manhattan Bank now argues that the small industrial boiler market will be lost to coal for good if oil prices stick at $30 per barrel or less(85). Other doubts remain about the extent to which there will be enough growth in the global electricity sector to sustain a major increase in overall coal demand, and there are also questions about whether the world steel industry has now lost so much ground that supplies of coking coal must be counted as an extra source of competition to traditional supplies of steam coal.

Symptomatic of the new wave of doubts about coal's future is the recent MITI projection that Japan's demand for coal until 1995 will be some 30 per cent below its previous estimates. This will still leave room for some increase in coal imports, but the signs are that these will not have reached the previous projection for 1990 ten years later(86). What is even more significant is that MITI is projecting that coal's penetration of the Japanese energy market may actually decline from the 18.5 per cent achieved in 1982 - a trend which would run counter to recent conventional wisdom about what a desirable energy policy should be. The softening of world oil prices has clearly led at least this key energy consumer to downgrade the future role for coal. Taking this development into account with the general run of pessimistic developments affecting coal, it would appear that it will be the lower projections listed in Table 6 which are more likely to be achieved.

However, although the balance of evidence shows that the volume of internationally traded coal is more likely than not to fall into the 200-300 million tonnes region in the year 2000, it is important to note that bodies like the Chase Manhattan Bank have still come out in 1983 with a 604 million tonne projection (which will include some coking coal). One should also not forget that such a figure will represent less than 10 per cent of world coal consumption, and that this coal trading will still be very much a marginal activity potentially vulnerable to major swings in its fortunes as the various indigenous supply-demand relationships develop. Finally, whether demand for traded steam coal is 200 or 600 million tonnes, it is worth noting that the British (particularly CEGB) demand for coal is quite substantial in relation to it. There is therefore some substance in the argument that, whatever the underlying marginal costs involved at any given demand level, the emergence of the UK as an importer of coal would give a significant psychological upward push to any given price level.

On balance, it looks increasingly as though steam coal trade in the year 2000 will lie somewhere in the 250-350 million tonnes range, with the possibility that it could be as high as 600 million. However, conceding that there is some uncertainty about the level of such demand, there is much greater uncertainty about what any demand forecast will actually mean in terms of international prices. As indicated in Figure 4, the lower figure of 250 million tonnes produces estimates of marginal supply costs which vary between \$50 and \$90 per tonne. At the upper end of demand forecasts, the variation is much greater, ranging between \$70 and \$130 per tonne.

It is possible to narrow the range of such forecasts if one is willing to dismiss the low-price argument as put forward by BP to the sub-committee of the House of Lords European Communities Committee(87). The CEGB certainly does, arguing that oil companies like BP are inexperienced in the coal industry and that they are anyway at a stage where they are trying to buy a market share(88). There is some supporting evidence for the CEGB pessimism about future marginal costs from the way the Exxon venture in Colombia is increasingly looking a high-cost one(89). On balance, the low-cost case seems to need some strengthening, because the high-cost arguments of the CEGB (and, to a lesser extent, the NCB) rest on more publicly exposed supporting evidence. The present author is unaware of any exponent of the low-cost supply curve who has backed this argument with a country-by-country analysis of the future availability of differing quantities of exportable coal at various prices. The WOCOL study is specifically included in this observation. Useful as that study has proved in alerting a world readership to the possibilities of the coal industry, it is not helpful in the construction of a supportable international supply curve (South Africa is given two pages of detailed analysis: Colombia one-and-a-third). This is not to say that the low-cost proponents will not prove correct, but it is striking that subsequent analysis which supplements WOCOL(90) tends to be rather less optimistic about the cost prospects of key exporters.

The result of all this is a degree of uncertainty which will make planning for the coal industry particularly difficult. The future of the international coal price is absolutely key to any confident decision-making about British coal. However, the combination of uncertainties about the level of future demand and what the marginal costs of supply will be for any given demand level, means that one can only give a price range varying between some $60 and $130 per tonne. This study's core assumption that the world economy will grow by 1.5 per cent per annum to the year 2000 and that the oil price will reach $43 per barrel suggests that demand will fall more toward the 300 million tonne level, and that the marginal cost will fall somewhere within the NCB and 'oil company' cases - i.e. around $70 per tonne. However, the range of uncertainties around this estimate is extremely wide.

British demand. The final element of the equation is what actual demand there will be for coal in the UK. Low demand will disappoint the British industry but will, under most scenarios, at least mean that what coal is needed can be economically supplied from British pits. As demand rises, so will the marginal British costs leading to circumstances where imports may be the most economic response.

The estimate made in this study's base case (the Department of Energy's case BL)(91) is for coal demand to be 116.2 million tonnes in the year 2000. As can be seen from Table 9, this is well down the list of recent forecasts of British coal demand and certainly well below the forecasts of the late 1970s when demand as high as 200 million tonnes was contemplated. However, there are industrialists, and academic analysts like Colin Robinson, who argue that demand could be well below even the estimates of the base case on the grounds that industrial demand will not be as high as once assumed; that high technology uses of coal like liquefaction and gasification are now unlikely to be significant during the timing of this study; and that CEGB demand will prove less buoyant than the Board is currently willing to admit. Following such analysis, it is possible to conceive of situations in which British demand could drop to as low as 70 million tonnes. This may sound extreme, but it is noticeable that the mainstream forecasts are still being edged downwards, with knowledgeable observers now starting to concede that demand in the year 2000 is most unlikely to be more than 110 million tonnes.

Again, even this figure is being challenged by some industrialists, who argue that demand may actually fall to the 70-90 million tonnes region (after deducting the requirements of coke ovens). This pessimism would depend on electricity demand continuing to climb at rates below those seen in, say, the 1950s and 1960s, and on the effect of the completion of nuclear reactors currently under construction on the demand for coal by the CEGB (each nuclear plant replaces roughly 4 million tonnes/year of coal). Add in some further pessimism about industrial demand, based on the work done recently on the age structure of industrial boilers(92) and one can conceive of the possibility that demand may well be below the 100 million tonne region in the year 2000.

However, this study was not intended to undertake a definitive analysis of future British coal demand; it will therefore do no more than note that there are serious voices both within and outside industry who reckon that demand estimates of British coal will need further downward revision. For the moment, let us stick with the Department of Energy's conservative estimate of 116 million tonnes at the turn of the century. Such a figure would be well within the British industry's potential. Assuming that there will be some drastic pruning of today's higher-cost pits, then it would not take a particularly ambitious investment programme to maintain capacity at such a level. Should demand run higher, however, (say above the 130 million tonnes level), then there could well be problems, for it is by no means clear that the NCB has sufficient deposits identified and ready to be given planning permission to raise capacity much above that level by 2000, and even if there were, there would be serious questions about the costs at which this new capacity would come in.

For any given supply curve, the British industry is not going to be massively sensitive to a shortfall in demand of some 20-30 million tonnes. All this would do is lower the NCB's average costs, providing it has not over-invested to supply a higher level of demand. A rule of thumb suggests that a swing of some 35 million tonnes would raise or lower the NCB's marginal costs by some £10 per tonne, which is considerably below some of the cost fluctuations involved in other scenarios discussed in this study. Certainly, providing the British industry does not maintain a massive surplus of capacity, fluctuations in the UK demand for coal are going to be much less important to the general economic case affecting imports than potential fluctuations in the international demand for traded coal.

The sensitivities summarised. This study's base case, therefore, concludes that the British industry would be totally self-sufficient, on its assumptions that:

- UK demand is 116 million tonnes
- there is no real escalation in British production costs
- the £:$ ratio is 1:1.50
- the demand for internationally traded coal is 300 million tonnes

- and the marginal cif cost at that level of demand is around $65 per tonne (about half-way between the oil company and NCB cases).

At that international price, the British industry could competitively supply some 130 million tonnes, should sufficient demand be there.

As has been mentioned several times, this base case rests on a fairly arbitrary, but far from impossible, set of assumptions. What matters is not whether the base case is likely, but the impact of changed assumptions on it, since it is the analysis of such variations which allows readers to make up their own minds about the likely outcome of the variant they find most convincing. To assist the debate, we now look at the circumstances which would produce a situation in which the UK would be in rough balance, and then at those which would produce a 30 million tonne demand for imports.

Variants which (individually) would make imports marginally competitive

- British demand rises to around 135 million tonnes
- British costs rise 1 per cent per annum in real terms
- £1:$1.90 sterling-dollar exchange rate
- international demand falls to around 200 million tonnes
- the international supply curve is virtually flat (i.e. the oil industry case)
- transport costs are reduced by $10 per tonne.

Variants which (individually) would make 30 million tonnes of foreign coal economic to import

- UK demand rises to around 165 million tonnes
- British costs rise some 2 per cent per annum in real terms
- transport costs are reduced by $20 per tonne
- (£1:$2 exchange rate would produce some 10 millions of imports)

These two sets of variants suggest that the UK could well find itself with an economic case for importing coal, but that there is no simple way of determining what the likely import pressures will be by the turn of the century. For one thing, British coal could obtain further protection from some combination of lower-than-predicted demand, a fall in sterling's rate against the dollar, higher-than-expected international demand for coal, and the outturn that marginal costs in the international arena are more along the lines of the CEGB's case. These are all perfectly plausible developments and should not be dismissed out of hand. On the other hand, there is certainly nothing implausible about British production costs rising by 2 per cent per annum in real terms, and the reduction of transport costs by $20 per tonne could come through the switch to larger bulk carriers, the development of appropriately-sized import terminals in the UK, and the steady relocation of coal-burning plants around these terminals.

On balance, though, it looks as though an adequately managed British coal industry should be able to keep imports at bay, providing the international coal price moves upwards toward the $90 per tonne region. It is a matter of serious debate, however, whether the international price really will reach such a level. If it stays more toward the $60 region, then the British industry will become very vulnerable to a variety of factors, many of which are not under its control. Some factors, such as developments in transport, will work inexorably to cheapen the cost of imports. Others, such as currency relationships, could work either way. The one factor which can be heavily influenced by good management is the containment of British production costs. Should these fail to fall below the recent long-term trend of 1.5 per cent annual increase, then an industry faced with $60 per tonne imported coal will come under heavy pressure from imports at its margin.

Policy issues

Given this tentative conclusion that, under many scenarios, a good part of the British industry could be competitive with imports, it is tempting to go a step further and argue that the UK should simply adopt a policy of self-sufficiency in coal, as a contribution to both employment policy and energy security. Such arguments need careful scrutiny - at least in part, because it is by no means guaranteed that the industry will be competitive. It has demonstrated an ability to run up large deficits, thus suggesting that there may well be a substantial price tag to such apparently unexceptionable policy goals.

Certainly, one should not too easily forget the economic arguments for letting the British coal industry fight imports without any particular government intervention. After all, British manufacturing industry has enough problems on its plate without also having to pay relatively high energy prices. In these circumstances, imports can serve as a useful check on the coal industry's apparent inability so far to control its costs rigorously. Banning them gives it a semi-monopolistic hold on parts of the electricity-generation industry, with a subsequent knock-on effect throughout the rest of the economy. Protecting employment in the pits may well merely transfer unemployment pressures elsewhere.

Against this economic rationale, there is the counter argument that the most pressing economic need at the moment is to restore the NCB's finances, and that this will require a quite extensive programme of pit closures and, undoubtedly, labour shedding. It would therefore be both tempting fate and somewhat inhumane to face the miners simultaneously with the need to cut capacity to bring it in line with demand, while also arguing that the CEGB should be free to bring in imported coal again. There is a strong case for fighting such battles one at a time.

A further reason for keeping coal prices competitive within Britain is the desirability on energy security grounds of limiting the country's use of oil. This may not seem immediately important to the UK as long as it is well supplied with indigenous oil, gas and coal. But the global impact of the oil price hikes of the 1970s was sufficient to show that no country is an island in energy matters. However self-sufficient Britain may be in energy, it will still suffer from oil-related economic disruption. The logical consequence of this argument is to ensure that British coal is as competitive as possible with alternative energy sources, particularly oil. Coal imports will help keep the British industry on its toes and, in a different sense of the security argument, will also provide some degree of flexibility should a body like the National Union of Mineworkers try again to hold the country to ransom. Imports can be important for consumers faced with a monopoly. Energy security does not just mean freedom from a dependence on foreign supplies.

In fact, the international coal industry does throw up a few security problems in its own right. The collapse of the Polish coal industry in 1980 was just as dramatic as the collapse of the Iranian oil industry in the last days of the Shah, and there is always the possibility that white control in South Africa will one day collapse, similarly plunging an increasingly important international supplier into chaos. Likewise, both the Australian and US union scenes are tempestuous ones and importers from these sources would clearly need to assume the possibility of lengthy labour disputes.

However, to point to such risks is not to argue that the only counter is total reliance on British coal. There could well be much cheaper ways of ensuring that security needs are catered for. At the very least, importers could be expected to spread their purchases (and risks) between two or three of the lead exporters. The speed with which the USA moved to fill Polish markets in 1980/1 is a reminder that the oil industry is not alone in knowing how to redirect energy flows in times of crisis. A further approach (which is also found in the oil industry) is to build stockpiles, not as a politically easy way of disposing of surplus production in times of glut, but as a positive contribution to energy security. Such policies will normally be cheaper and easier than maintaining some kind of 'surge' capacity in the British industry to make up for international shortfalls - and this assumes that 'surge' capacity makes any kind of sense in the British context. The idea of 'mothballing' some British pits has little attraction, and it would also be quite difficult to persuade British miners to work extra shifts should one or more of their foreign competitors run into trouble.

So, taking a number of these considerations into account, British policy-makers have to plan for an industry which has too much capacity and is marginally vulnerable to imports. A drastic pruning of the labour force will be politically difficult, and there is no clear sense as to where the cuts should stop. On the one hand, forecasts of demand for coal in the UK are still falling. On the other hand, for any given level of demand, it is impossible to predict with confidence what proportion of the British industry will be competitive with imports. Politicians and managers are thus faced with precisely the kind of decision they hate - one in which there are still extremely sensitive political and social questions to be settled with a hitherto powerful trade union over issues which leave the relevant experts more than usually divided. What should policy be?

Probably, as we have said, the first priority must be to get the NCB finances on to a proper footing. This will inevitably mean pit closures, and a transitional protectionist policy could be justified if it made them politically easier to achieve.

However, once that short-term goal is achieved, the problem of imports will not go away - but there is no guarantee that they will be competitive with the British industry. For the reasons argued above, it is important that no decisions be taken which may block imports should they be needed. This means that the NCB should be encouraged to be relatively cautious about its demand estimates when planning future capacity levels. It is probably more expensive for the economy to support surplus NCB capacity than to import a similar quantity of coal should the NCB underestimate demand. At the same time, the CEGB should also be encouraged to keep its options open.

Policy should reflect the benefits to be gained from the greater competitive pressure on domestic costs which imports can exert. Planning delays should be avoided if the demand for a deep-water terminal builds up. However, one can argue that governments should think more ambitiously. If it is legitimate to subsidise companies to convert oil and gas boilers to coal, then it should also be legitimate to provide some incentive to anyone seeking to develop such a terminal. A general knowledge that coal imports could be handled in substantial volumes if economics so dictate would have a symbolic impact throughout the energy-intensive parts of the British economy. It is precisely in circumstances like these that a government should take a longer view than an individual corporation (in this case, probably the CEGB?) which has to decide on its own narrow considerations.

In addition to the contribution which imports could make to existing demand patterns, there is the further question of 'forcing' coal combustion technologies to the point where commercial interests can immediately adopt them once the underlying economics of coal are right. Clearly, the immediate need for gasification and liquefaction technologies has been reduced, but there is an intermediate range of developments primarily concerned with coal combustion which continues to deserve official support so long as the increased use of coal puts further downward pressure on the price of oil. Similarly, continued pollution worries about the burning of coal suggest that government should put some effort into environmental research: it is pointless to try to maximise coal production if potential demand is constrained by such worries - or if indeed these fears turn out to have been justified.

At the end of the day, though, this study is primarily concerned with imports, and the message which comes over clearly is that there are still major uncertainties about forecasting the international price of coal. Since this will heavily affect the future size and profitability of the British coal industry, it is particularly important that policy-makers clarify the disagreements between the various arguments. In the meantime, planning has to occur in an uncertain environment and this means planning flexibility. It should certainly be made clear by government to the CEGB that, once the immediate political problem of slimming down the coal industry is over, it should be free to re-start importing if that is economic. There is no reason why the CEGB's ability to import should not be phased over several years to limit any shock to the British coal industry. And in the background, the government should encourage thinking about a deep-water import terminal, giving financial backing if that will encourage a developer to take a moderately daring risk.

With luck, though, the bulk of the British coal industry may be fully competitive with imports. But 'luck' is the operative word, and what this study has tried to do is to indicate where luck comes into the equation.

EPILOGUE

Since this report was first published in May 1984 two things have changed. First, the collapse of the coal strike has diminished, but by no means eliminated, the power of the National Union of Mineworkers within the UK industry. Second, the pound has been down to within ten cents of the dollar. However, for reasons given in the partially revised text of this paper, the present exchange rate does not provide a sound basis for consideration of the future of the industry. On the other hand, the present situation does provide the nation, and the National Coal Board itself, with the first opportunity since nationalisation in 1946 for a fundamental reappraisal of the role of indigenous coal in UK energy.

It is fashionable to decry planning, and to dismiss as utopian if not positively dangerous the idea of energy policy. Those who do so assert that all that is required is to be flexible and to respond to the market. In coal as elsewhere, however, the decision to invest involves a view on what the world will be like when the new investment comes into production, while no manager can afford to ignore the inheritance of human and physical equipment which exists in his industry. In that sense, a business must plan. This is not to say that change in the criteria for investment, or in management style and organisation, is impossible or undesirable. What it does suggest, amongst other things, is that privatisation of the NCB's inheritance is unlikely to succeed if only because of the certainty that it would encounter impossible management problems and would be reversed by a future government of different political views.

So what practical action can now be recommended to government and to the management of the NCB recognising that outsiders cannot know the limits to the degree of manoeuvre that is open to them?

First, as to government. No government can abdicate responsibility for the framework within which a State industry will operate, though it may be accepted as axiomatic that the less it interferes in the day to day management the better. In fact, in order to allow the managerial programme of action suggested below, government would need to consider taking the following action, some of which may require legislation:

a) lift the (de facto) prohibition on the CEGB importing foreign coal, but still expect it to buy at least two thirds of its coal requirement (pre-strike totalling 75mt a year) from the NCB (instead of 100 per cent at present). Encourage the CEGB or private concerns to build some power stations at major ports, designed to run on imported coal;

b) provide capital as now planned for new NCB investment and confirm the pledges made by the Secretary of State for Energy during the strike that there will be no compulsory redundancies; difficult though this may be where pits have been irretrievably damaged by the strike;

c) provide government funds to cushion the transition including capital for alternative employment in areas of mine closures; (It needs however to be recognised that this will involve the taxpayer in provision for one part of the community, consisting of less than 0.5 per cent of the workforce, on a scale unprecedented for any other part. The electorate has a right to insist that this not be an open-ended commitment.)

d) permit the management of the NCB to operate under normal commercial criteria, and not distort these by arbitrary Treasury accounting rules. This almost certainly involves a capital restructuring;

e) then let the NCB get on with the job.

It would be consistent with the above to offer the following suggestions to the NCB itself:

a) the opportunity exists to revitalise the management, many of whom, having worked in a declining industry for 25 years, may not have either the flexibility to entertain new ideas or the ability to accept accountability for decentralised management. This means early retirements, generous compensation, and replacement from the young or from outside. At the other end of the workforce improve the relative wages for deputies and faceworkers;

b) undertake a blitz on overheads with the aid of outside consultants and sell off redundant property;

c) put in hand, with the aid of outside consultants and the appropriate consideration within the industry, a pit by pit review, in order to determine which pits now make a positive contribution to overheads, against the following criteria:

   i) all past investment to be ignored. These sunk costs may never be recovered, and they are irrelevant to the decision whether to continue production or not (hence the need to revise the financial objectives imposed by the Treasury);

   ii) assume that coal will fetch the same netback price at the pit head in £ of 1983 that it did in that year, i.e. assume a constant real price and that any increase in real receipts will be offset by increases in real costs. (Nobody knows whether a price set by competition at the margin with the internat ional market will be higher or lower than in 1983; no assumption that it will be higher is sufficiently robust for planning purposes);

iii) provide new investment (including major repairs necessitated by the strike) only where it will increase the positive contribution to overheads sufficiently to provide a commercial return on that new investment (say 10 per cent in money-of-the-day or 5 per cent real). In other words, the return from the new investment must at least cover the long-term cost of borrowing in the market;

d) close the pits which are found not to make a positive contribution to overheads. The costs of redundancy, relocation and alternative employment payments should be ignored for decision purposes. Even if in the short term they exceed the savings from closure, in the long term their retention will once again shackle the NCB with an inheritance of uneconomic and unmanageable assets;

e) continue the investment in Selby and Belvoir but delay any further decisions on new mines until experience has shown that costs can be contained in these two new pits and that there is a probability of a market for additional production;

f) meantime do not assume for planning purposes that the UK market for coal in 2000 will exceed 100mt (the 1983 level), or that the NCB's market share will be more than 85mt (it would drop to the 50mt implied in the proposed CEGB guarantee, unless costs both in existing and new pits are kept below the delivered price of imports to inland sites);

g) assume no exports, other than to dispose of unplanned surpluses that cannot better be put to stock. Extra production for export must by definition be at the marginal (highest) cost of the industry. This, combined with the cost of transport to seaboard and shipment and unshipment (i.e. the converse of the logistic situation that will continue to protect most UK coal from foreign competition), makes it unrealistic to suppose that UK coal can compete on the Continent with that produced in Australia, South Africa, Colombia, Poland or even the United States and shipped in large carriers to bulk import terminals. Given the slower-than-expected growth of the international coal trade, and the spare capacity at current cost that exists in most of these countries, there is no justification for any assumption that delivered prices in real terms in North West Europe will rise above present levels. Exceptions may exist for small quantities of special qualities or small shipments to small ports in, for example, Scandinavia;

h) concentrate both market research and technical research on the consumer; for example, on improved combustion and delivery systems, and on services that would encourage the commercial and even the domestic consumer to burn coal. Examples are technologies such as pulverised coal mixed with water or oil which would facilitate alternative firing in industry and electricity generation, or fluid bed combustion which promises to be both more efficient and environmentally acceptable as a means of electricity generation. No technical effort or money should be wasted on developing known technologies whose commercial application is 20 years or more away, such as existing techniques for coal liquefaction. These technologies can be bought in if and when the time comes, from those who have developed them overseas in countries where the economics or the politics are different. However the NCB should continue to sponsor research to provide new understanding of the basic science of coal, for which there might be a commercial prize in the future.

Given an industrial approach on the above lines, there seems no reason why the NCB's should not become by 1990 a viable operation selling no less coal than in 1983, and making a corresponding contribution to the national wealth as well as to the prosperity of those who work in the industry.

16 April 1985

NOTES

(1) Monopolies and Mergers Commission, National Coal Board: A Report on Efficiency and Costs in the Development, Production and Supply of Coal by the NCB, 2 Vols, London: HMSO, 1983.

(2) This study concentrates almost entirely on steam coal, that is, coal which is primarily used for electricity generation and various kinds of heating. It does not deal with coking (or metallurgical) coal which is primarily used in the steel industry.

(3) This is a National Coal Board estimate, which is treated with much suspicion by the British Geological Survey, amongst others.

(4) WOCOL study. Robert P. Greene and J. Michael Gallagher, Future Coal Prospects: Country and Regional Assessments: Report of the World Coal Study, Cambridge, Mass: Ballinger, 1980, p.394.

(5) In 1981 underground production costs (both capital and operating) of South African underground mines were one sixth the average cost of British coal. Australian underground coal was one quarter as costly, while Canadian and US mines were around 60 per cent of British costs. Calculated from Ray Long, Coal and the Costs of Energy, Working Paper for IEA Coal Research, London, 1983, p.4.

(6) Neil K. Buxton, The Economic Development of the British Coal Industry, London: Batsford, 1978, p.166.

(7) See studies such as Buxton, op.cit.; Peter James, The Future of Coal, London: Macmillan, 1982; Gerald Manners, Coal in Britain; An Uncertain Future, London: Allen and Unwin, 1981; Tony Hall, King Coal: Miners, Coal and Britain's Industrial Future, Harmondsworth, Middlesex: Penguin, 1981; Colin Robinson and Eileen Marshall, What Future for British Coal Policy?, Surrey energy economics discussion paper, No.14, Guildford: Surrey University, 1983.

(8) House of Commons, Energy Committee, Pit Closures, 2nd report, session 1982-83, HC135, London: HMSO, 1982.

(9) Buxton, op.cit., p.239.

(10) National Coal Board, Plan for Coal, London: NCB, 1974.

(11) Department of Energy, Green Paper on Energy Policy, Cmnd 7101, London: HMSO, 1978.

(12) Monopolies and Mergers Commission, Central Electricity Generating Board: a Report on the Operation by the Board of its System for the Generation and Supply of Electricity in Bulk, London: HMSO, 1981.

(13) Monopolies and Mergers Commission, National Coal Board, op.cit., Vol. 1, p.364.

(14) Ibid., p.370.

(15) Ibid., p.60.

(16) Ibid., p.65.

(17) Ibid., p.368.

(18) Ibid., p.368.

(19) Ibid., pp.365-6.

(20) Ibid., p.76.

(21) CEGB P4 (by F.P. Jenkin), On: the Need for Sizewell 'B', Proof of Evidence (P4) to Sizewell 'B' Power Station Public Inquiry, London: CEGB, 1982, p.78.

(22) Ibid., p.98.

(23) This is the CEGB's Case C (see Appendix).

(24) CEGB P4, op.cit., p.78.

(25) CEGB P6 (by P.R. Hughes), On: Fossil Fuel Supplies, Proof of Evidence (P6) to Sizewell 'B' Power Station Public Inquiry, London: CEGB, 1982, p.75.

(26) Ibid., p.89. It should be borne in mind that all prices, costs and coal measurements have been standardised as much as possible:

- a 1983 dollar is used throughout (£1:$1.50)
- a tonne of coal is assumed to be 29.3 Giga-Joules.

This has necessitated some fairly heroic assumptions when recalculating recent forecasts made at a time of quite rapid currency movements. In general, it is assumed that 1981/2 estimates of future South African and Australian costs should be downgraded in line with the movements of their currencies against the dollar. It has been assumed that the devaluation of the Colombian peso will not have significantly improved the economics of that country's coal export potential. It is also accepted that a 29.3 GJ tonne of coal is not often found in the steam coal trade, but it provides a numeraire with which a number of forecasters are happy.

(27) Ibid., p.88.

(28) CEGB P4, op.cit., p.68.

(29) CEGB P6, op.cit., pp.65-7.

(30) Monopolies and Mergers Commission, National Coal Board, op.cit., Vol. 1, p.368.

(31) Manners, op.cit., p.16.

(32) CEGB P6, op.cit., p.47.

(33) M.J. Allen, 'Some geological criteria for investment in coal mining', The Mining Engineer, (April 1982), pp.583-8.

(34) Monopolies and Mergers Commission, National Coal Board, op.cit., Vol. 1, p.247.

(35) Ibid., p.234.

(36) CEGB P6, op.cit., p.60.

(37) Ibid., p.73.

(38) Monopolies and Mergers Commission, National Coal Board, op.cit., Vol. 1, pp.228-9.

(39) WOCOL, Coal: Bridge to the Future: Report of the World Coal Study, Cambridge, Mass: Ballinger, 1980; Greene and Gallagher, op.cit.

(40) CEGB P6, op.cit., p.88; CEGB/P/6 (add 5), Further information provided by Mr P.R. Hughes, London: CEGB, 1983, p.3.

(41) Ray Long, Constraints on International Trade in Coal, EAS Report No. G3/83, London: IEA Coal Research, 1982.

(42) Frank Pecchioli, 'World trade in coal', address to Royal Institute of International Affairs, 23 February 1983 (mimeo).

(43) CEGB/P/6 (add 5), op.cit., p.3.

(44) Robertson Research: Bill Fischer, cited International Coal Report, January 1985, pp.2-4.

(45) CEGB P4, op.cit., CEGB P6, op.cit., CEGB P5 (by C.H. Davies), On: Scenarios and Electricity Demand, Proof of Evidence to Sizewell 'B' Power Station Public Inquiry, London: CEGB, 1982. NCB/P/1 (by M.J. Parker), On: Coal Price Prospects and Availability of Coal in the UK Power Generation Market, NCB, Proof of Evidence, London: NCB, 1983.

(46) Shell: see Pecchioli, op.cit. British Petroleum, memorandum submitted to the House of Lords, European Communities Committee (sub-committee F) on 10 March 1983.

(47) BP, op.cit., p.60. Steenblik's views reported in The Guardian, 5 November 1983. Richard Eden and Nigel Evans, Electricity's Contribution to UK Self-Sufficiency, BIJEPP Paper No.11, London: Heinemann, 1984.

(48) CEGB P6 op.cit., Table 8, p.89: CEGB/P/6 (add 5) op.cit., p.5.

(49) CEGB P4, op.cit., p.45.

(50) NCB/P/1 (add 3), Written replies by the National Coal Board to questions raised on Mr Parker's evidence to the Inquiry (NCB/P/1), additional evidence to the Sizewell 'B' Power Station Public Inquiry, London: NCB, July 1983, p.27.

(51) Ibid., p.26 (these figures have been converted to the autumn 1983 dollar-sterling exchange rate).

(52) Hugh Mellanby Lee, The Future Cost and Availability of Thermal Coal Exports from South Africa, Economic Assessment Paper, working paper, London: NCB (IEA Services), 1982.

(53) Ibid., p.21.

(54) Long, Constraints on International Trade, op.cit., p.47. Richards Bay is currently expanding its capacity to 44 million tonnes, while the authorities have extended export licenses to

cover 80 million tonnes (memorandum submitted by BP to the House of Lords, op.cit.

(55) Lee, op.cit., p.41. P.R. Hughes in cross examination on day 79 of the Sizewell Inquiry, p.17 of the official transcript (17 June 1983).

(56) Pecchioli, op.cit. From this point on, all costs are given cif in North West Europe in order to ease the analysis.

(57) CEGB/P/6 (add 6), Further information provided by Mr P.R. Hughes: comparison with Chase Econometrics Studies, evidence to Sizewell 'B' Inquiry, London: CEGB, July 1983.

(58) P.R. Hughes in oral evidence to the Sizewell Inquiry (17 June 1983), pp.15-17 of the official transcript. CEGB/P/6 (add 6), op.cit.

(59) Lee, op.cit., p.43.

(60) Pecchioli, op.cit.

(61) Ibid., p.7.

(62) Long, Constraints, op.cit., p.17.

(63) NCB/P/1 (add 1), Written replies by M.J. Parker to elucidatory questions raised by the CEGB and Professor Foster on NCB/P/1, evidence to Sizewell 'B' Inquiry, London: NCB, April 1983, Tables 7-8.

(64) Ibid., para. 5(iii).

(65) Energy Economist, September 1983, pp.11-12.

(66) NCB/P/1 (add 1), op.cit.

(67) Energy Economist, September 1983, pp.11-12.

(68) Ibid., p.12.

(69) Long, Constraints, op.cit., p.33.

(70) Energy Economist, September 1983, pp.11-12.

(71) NCB/P/1 (add 1) op.cit., Table 2.

(72) Ibid., Tables 7 and 8.

(73) See Coaltrans 82, The 2nd International Coal Trade, Transportation and Handling Conference, Worcester Park, Surrey: C S Publications Ltd, 1982.

(74) Ray Long, op.cit., Hugh Mellanby Lee, op.cit., Hugh Mellanby Lee, The Future Cost and Availability of Thermal Coal Exports from Australia, working paper No.55, Economic Assessment Service, London: NCB (IEA Services), 1982.

(75) See Note to Figure 3 for assumptions made.

(76) CEGB P6, p.73.

(77) Colombia is currently looking less attractive than it once did; a report in the October 1983 Financial Times Energy Economist (p.5) suggests that the Exxon El Cerrejon joint venture was planned on the assumption of $89 per tonne fob prices in 1986. Be that as it may, the investment going into this project is developing a coal-exporting infrastructure which should keep the cost of future investments under control.

(78) CEGB P6, op.cit., pp.114-5.

(79) Resource Policy Centre, Transportation of US Export Coal through Eastern Ports, prepared for the US Department of Energy, DSD 456, Hanover, New Hampshire: Dartmouth College, 1982, p.36.

(80) US Department of Energy, Report on Potential for Cost Reductions in Inland Transportation of US Coal Exports, Washington DC: Department of Energy, 1983.

(81) Ibid.

(82) Monopolies and Mergers Commission, National Coal Board, op.cit., Vol.1, p.74.

(83) CEGB/P/6 Further information provided by Mr P.R. Hughes following his cross-examination..., London: CEGB, June 1983, p.5.

(84) Liverpool on the West coast is occasionally mentioned as a possible site. It has the advantage of being an under-employed existing port with a coal-fired station (Fiddler's Ferry) extremely close to it. The Midland belt of similar stations are all about 100 miles away. This is, however, at least double the average distance of these stations from the East coast, and a terminal there would be in a far better position to supply other coastal-based stations on the East coast in Scotland and the Thames than would Liverpool.

(85) Financial Times Energy Economist, (April 1983), p.14.

(86) Ibid., October 1983, pp.1-2.

(87) BP, op.cit.

(88) P.R. Hughes in oral evidence to Sizewell Inquiry, 17 June 1983, p.33 of official transcript.

(89) Financial Times Energy Economist, October 1983, pp.5-6.

(90) Such as Long, Constraints, op.cit.

(91) Department of Energy, op.cit., p.A44.

(92) John Chesshire and Mike Robson, UK Industrial Energy Demand: Economic and Technical Change in the Steam Boiler Stock, SPRU Occasional Paper No.19, Brighton, Science Policy Research Unit, 1983.

Appendix: Scenarios

It is currently fashionable in planning circles to work off a number of scenarios. This is healthy in one sense, but still begs the question of what the planners think is most likely. The present study is based on the relatively conservative scenario 'BL' which is one which the Department of Energy presented to the Sizewell Inquiry(1). This makes three basic assumptions:

- that world oil prices will stay low, rising to no more than $43 per barrel in the year 2000
- that British GDP growth stays quite moderate at 1.5 per cent per annum
- and that industrial growth remains low.

Toward the end of this study, a couple of further assumptions are added to provide a scenario of how the British industry's competitiveness with the international coal industry could develop:

- British coal costs are assumed to stay constant in real terms
- the pound-dollar ratio is taken as £1:$1.50.

Apart from the fact that the Department's BL scenario currently suits the relatively pessimistic view of most observers about the immediate future of the world (and British) economy, this scenario roughly approximates to the CEGB's scenario C in its Sizewell evidence(2). This assumes:

- an oil price of $50 per barrel in the year 2000
- British GDP growth of 1.0 per cent per annum
- UK industrial output growing at 0.5 per cent per annum.

Both the Department of Energy and the CEGB consider a number of other scenarios which imply both higher and lower energy demand, but rather than analyse each of these, this study considers variations round its own base case. On balance, the study assumes that the BL scenario is reasonably likely, with the assumptions about British coal competitiveness somewhat less so. However, the latter assumptions are made because they are relatively simple and thus help the analysis of variants around the base case they support. In other words, these cases are useful as pegs on which to hang a discussion. Doubtless, the real world will confound them.

NOTES

(1) Shell's scenario work is now quite well known. The British Government has also embraced this approach. See the 'Byatt Report' - Oil Prices in the Long Term: An Examination of Trends in Energy Supply and Demand and Their Implications for the Price of Oil, a study by an inter-departmental group of officials, London: Department of Energy, 1982. The BL scenario was presented in Department of Energy, Proof of Evidence for the Sizewell 'B' Public Inquiry, London: Department of Energy, 1982.

(2) CEGB/P/6 (add 3), op.cit., p.15. This is an addition to CEGB P6 (by P.R. Hughes), On: fossil fuel supplies, Proof of Evidence (P6) to Sizewell 'B' Power Station Public Inquiry, London: CEGB, 1982.

Table 1 UK coal industry statistics

	1947	1950	1955	1960	1965-66	1970-71	1975-76	1980-81	1981-82
Total output (mt)	200.0	219.8	225.2	197.8	185.7	144.7	125.8	126.6	124.3
Consumption (mt)	187.5	205.9	218.7	199.9	184.0	150.7	122.2	120.3	116.8
Imports (mt)	0.7	0.0	11.8	-	-	1.2	4.8	7.3	-
Exports (mt)	5.3	17.2	14.1	5.6	3.7	3.0	1.4	4.7	9.4
Manpower ('000)	703.9	690.8	698.7	602.2	455.7	287.2	247.1	229.8	218.5
Coal as % of UK primary inland energy consumption	90.8	89.6	85.4	73.7	61.8 (1965)	46.6 (1970)	36.9 (1975)	36.7 (1980)	37.4 (1981) (prov)

Source: Monopolies and Mergers Commission, The National Coal Board, op.cit., Vol.1, p.7.

Table 2 UK coal imports 1950-81 (mt)

1950	-
1960	-
1970	0.1
1973	1.6
1974	3.5
1975	5.0
1976	2.8
1977	2.4
1978	2.4
1979	4.4
1980	7.3
1981	4.4

Source: *Ibid.*, p.58.

Table 3 UK deep mines operating results 1972-3 to 1981-2

Year to March	Pro-duction (mt)	Disposals (mt)	Stocks (end yr) (mt)	Operating surplus (loss) no interest (£m)
1972-3	131.8	127.4	8.9	(59.9)
1973-4	98.7	102.1	5.3	(135.7)
1974-5	116.8	118.2	4.0	(35.7)
1975-6	114.5	108.8	9.7	(31.1)
1976-7	108.4	109.5	8.7	31.9
1977-8	106.2	105.9	8.8	0.3
1978-9	107.7	104.9	11.4	(26.1)
1979-80	108.6	110.0	10.1	(121.4)
1980-81	109.6	103.4	16.1	(106.5)
1981-2	108.2	103.9	20.4	(226.1)

Source: Ibid., p.27.

Table 4 Comparison of cost to the CEGB of NCB and imported coal of equivalent calorific value in pence per gigajoule (GJ) and pounds per tonne (situation in October 1982)

	Kingsnorth Tilbury West Thurrock		Didcot (Berks)		Ratcliffe (Notts)		Fiddlers Ferry (Cheshire)	
	p	£	p	£	p	£	p	£
Typical NCB coal	190	55.7	190	55.7	173	50.7	175	51.2
Imported coal								
(a) existing contracts	175	51.2	200	58.6	196	57.4	198	58.1
(b) New spot purchases	141	41.3	165	48.3	162	47.5	163	47.6

Assumptions: Average price under existing contracts (Continent)=$75/tonne (43.9). Present spot price (Continent)=$57/tonne (33.8). £1=$1.70.

On-costs from Continent: (a) to Thames 25p/GJ (£7.32); (b) to Didcot 50p/GJ (£14.64); (c) Ratcliffe 46p/GJ (£13.46); (d) Fiddlers Ferry 48p/GJ (£14.05)

Calorific value = 29.3 GJ/tonne (net) - note that the figures in the above table have been converted from a 25.5 GJ/tonne basis used in the original text.

On-costs to Thameside based on CEGB experience: costs to inland stations estimated.

Source: *Ibid.*, citing CEGB submission of October 1982.

Table 5 Real sterling/dollar exchange rate

	(March 1982 prices)	(Index 1975=100)
	£:$	
1960	1.26	84
1970	1.22	82
1980	2.21	147
1981	1.98	132
1982 (autumn)	1.72	115
2000		
CEGB A Case	1.60	107
CEGB B	1.06	71
CEGB C	1.33	89
CEGB D	1.60	107
CEGB E	1.77	118

Source: CEGB P5, On: Scenarios and Electricity Demand, Proof of Evidence to Sizewell 'B' Power Station Public Inquiry, London: CEGB, 1982, p.79.

Table 6 Comparison of world steam coal trade studies (mtce) (ranked according to year 2000 - all steam trade)

No.	STUDY	YR OF STUDY	1990 All steam	1990 Sea-borne	2000 All steam	2000 Sea-borne
1	WOCOL Case B. Bridge to the Future p.107, Tables 2-4, p.106, Figure 2-10	1980	260	-	680	-
2	CEGB High P6, p.88, Table 5	1982	160	131*	680	603a
3	NCB High/NCB/P1/Add 1 Para.4, Tables 2&4	1983	-	200	-	509
4	US Energy Information Admin. - High(b)	1982	306	-	567	-
5	US Energy Information Admin. - Central(b)	1982	271	-	501	-
6	US Energy Information Admin - Low(h)	1982	235	-	473	-
7	CEGB Middle P6, p.88, Table 5	1982	150	121*	380	359a
8	Chase Econometrics Base-line	1983	209	-	361	-
9	Chase Econometrics Variant 1 Midworld	1983	220	-	358	-
10	OECD - Steam Coal Prospects to 2000, p.144	1978	149	-	337	-
11	WOCOL Case A. Bridge to the Future. p.107, Table 2-4, p.106, Figure 2-10	1980	165	-	300	-
12	CEGB Low P6 p.88, Table 3	1982	140	116a	270	266a
13	NCB Low/NCB P1/(Add 1) Tables 1&3	1983	-	150	-	250
14	Johnson CPRE/P2REV./ p.40, Table 6a	1983	130	-	270	-
Additional reference						
(i)	BHP	1983	285c	-	-	-
(ii)	Shell	1983	-	240d	-	-

(a) Estimate based on assumed trading pattern.

(b) Prospects for Future World Coal Trade Document DOE/EIA/0363 - Dec. 1982, p.18, Table 12.

(c) Address by R.J. Burge (BHP) to Australian Coal Association Seminar 15 April 1983 - Figure taken from unnumbered diagram.

(d) F. Pecchioli - Paper to Royal Institute of International Affairs, Feb. 1983 - CPRE/P2 REV p.46.

Table 7 Rank and sulphur content of world coal resources

	Anthracite & Bituminous	Sub-bituminous Lignite	Peat	Sulphur Content (% weight) A&B	S&L
	bn tonnes				
Major exporters(a)					
Australia	25.4	33.9	-	1	1
Canada	1.7	4.3	0.5	1	1
China	99.0	0.1	-	n.a.	n.a.
Poland	27.0	12.0	-	1-3	1-3
S. Africa	25.3	-	-	1	-
USA	107.2	116.1	-	1-3	1
USSR	104.0	129.0	10.9	n.a.	-
Other major producers:					
Colombia	1.0	-	-	1	n.a.
German FR	24.0	35.2	0.9	0-3	n.a.
India	12.6	1.6	-	1.	1-3
Mexico	1.2	0.4	-	1-3	1-3
Spain	0.4	0.6	-	1-3	3
UK	45.0	-	-	1-3	n.a.
Rest of World					
Africa	7.2	0.2	-	1-3	1
Asia	2.3	3.4	-	n.a.	n.a.
E. Europe	3.1	53.2	-	1-3	1
Latin America	0.3	2.2	0.1	n.a.	n.a.
Oceania	-	0.2	0	n.a.	n.a.
W. Europe	1.1	1.7	3.4	1-3	1-3
World Total Recoverable	487.8	394.0	15.8		
World Total-in-Place	774.7	545.1	56,7		
World Total - Additional Resources	6161.4	5994.7	261.1		

(a) Figures for countries and regions refer to recoverable reserves as defined by the World Energy Conference as opposed to reserves in place or the reserve base. Definitions of recoverable reserves vary from country to country and recoverable reserves have recently been increased substantially for some countries (for example, S. Africa). Data for sulphur content vary widely, and may exceed the figures quoted.

Source: Ray Long, Constraints on International Trade in Coal, London: IEA Coal Research, Economic Assessment Service, 1982, p.22.

Table 8 Current UK port potential

Port	Max. vessel size ('000 dwt)
Ports currently used by the CEGB	
Birkenhead	15
Cardiff	23
Newport	33
Immingham	30
Thames & South East Power Stations	18
Scotland (Hunterston)	200
Ports irregularly used by the CEGB	
Port Talbot	100
Immingham Terminal	60
Jarrow	30
Swansea	30
Hull	25
Barry	15
Garston	12

Source: CEGB/P/6 (add 4) Further information provided by Mr P.R. Hughes, Sizewell Inquiry, London: CEGB, June 1983.

Table 9 Various estimates of British coal demand in year 2000 (mtce)

116.2	Dept. of Energy (Case BL)
200-150	WOCOL
200-135	NCB 1977
170	NCB 1979
142-113	NCB 1982
140-120	Dept. of Energy 1982
137	CEGB (Sizewell: Case C: low nuclear)
119	CEGB (Sizewell: Case C: medium nuclear)
115-80	Robinson 1979
114	CEGB (Sizewell: Case C: high nuclear)
110-75	Robinson and Marshall 1981

Dept. of Energy (BL)	Proof of Evidence for the Sizewell 'B' Public Inquiry, London: Dept. of Energy, 1982.
WOCOL	World Coal Study: Global Perspectives to 2000, London: NCB, 1980.
NCB 1977	Coal for the Future, London, NCB, 1977.
NCB 1979	Statement to the Commission on Energy and the Environment.
NCB 1982	Development Plan Review, London: NCB, 1982.
CEGB	P5 (by C.H. Davies), On: Scenarios and Electricity Demand, Proof of Evidence to Sizewell 'B' Power Station Public Inquiry, London, CEGB, 1982, (add 2), Table 10.
Robinson1979	Evidence to Vale of Belvoir Inquiry.
Robinson & Marshall 1981	What Futures for British Coal Policy, Surrey energy economics discussion paper No.14, Guildford: University of Surrey, 1983.

Figure 1 British operating costs 1981/82
(tonne £)

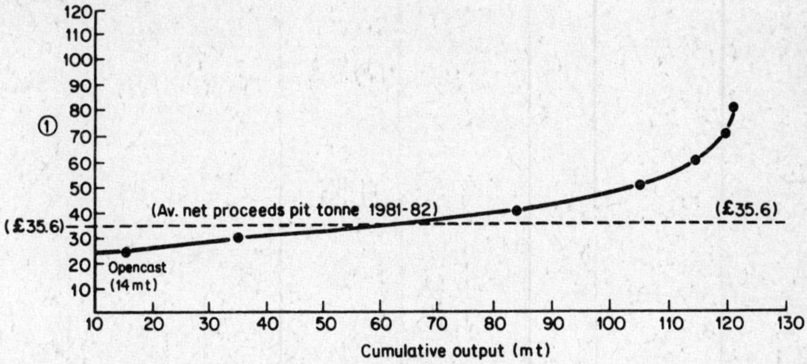

Constructed from pit-by-pit information in Monopolies and Mergers Commission, National Coal Board, op cit

① No Capital Costs – at minemouth

Figure 2 British operating costs in the year 2000
(tonne £)

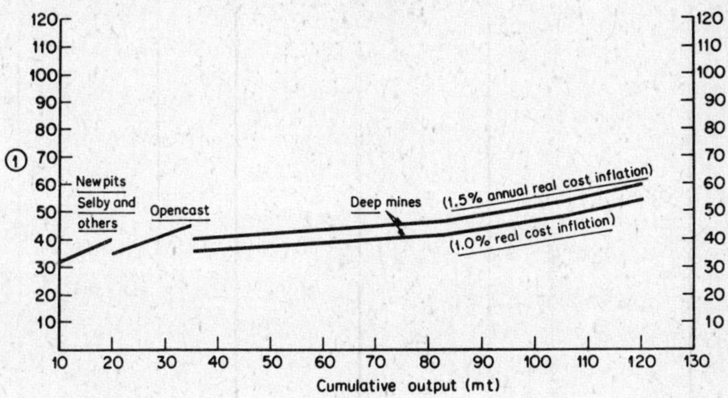

Source: Author's extrapolation from Figure 1

① Includes capital charges for new developments like Selby. Assumes existing deep mines are fully depreciated

Figure 3 Landed cost (cif) in N.W. Europe
(tonne $ autumn 83)

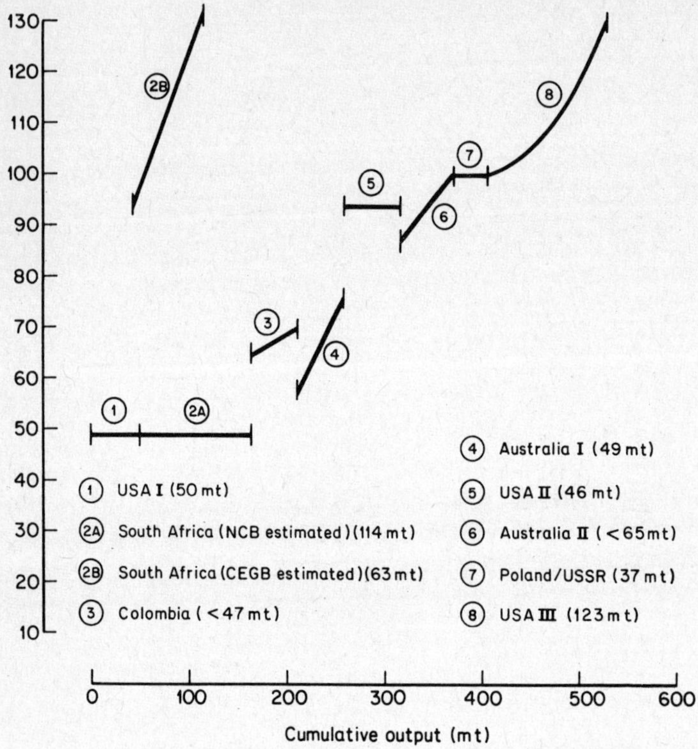

Sources primarily from:

- Ray Long, op cit
- CEGB/P/6 (ADD 6)
- Lee, op cit
- NCB/P/1 (ADD 1)

Adjustments have been made for currency movements since they made their estimates over 1981 and 1982. It is thus assumed that Australian and South African coal have improved their competitiveness against US coal to reflect the declines in their respective currencies, also that Colombian inflation will have roughly compensated for the decline of the peso against the dollar.

Figure 4 Alternative international supply curves
(landed costs in N.W. Europe: $)

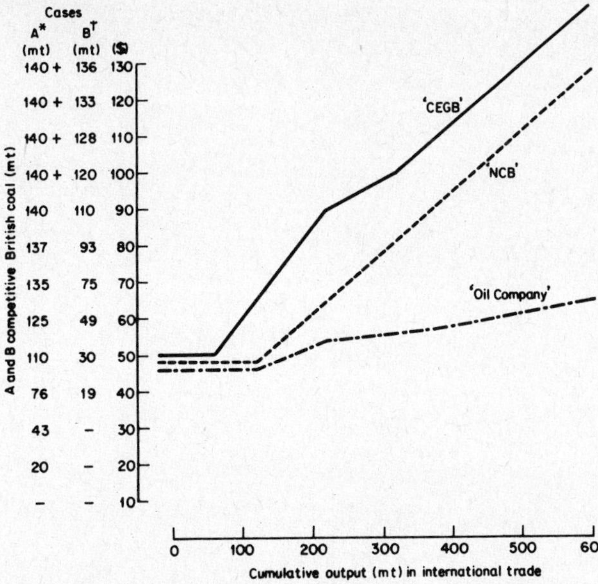

Case A* is favourable to the U.K. industry: the sterling/dollar rate stays at £1:$1.50: no real cost rises

Case B† is unfavourable: sterling strengthens to £1:$1.75: British coal costs increase at 2% p.a.

Sources: As for Figure 3

'Oil Company' curve is loosely based on assumptions expressed in presentations by Shell and British Petroleum: Frank Pecchioli 'World Trade in Coal', address to the Royal Institute of International Affairs, 23 Feb.'83 (mimeo) B.P. memorandum submitted to the House of Lords, European Committee (sub Committee F) on 10 March 1983

Note: assumptions

(i) Supply curve for British coal, as laid out in Fig 2 is broadly acceptable

(ii) Oncosts from ARA, as laid out in table 4, stay stable in real terms

(iii) Oncosts decrease in a linear fashion from £14 ($21) for the first tonne of demand for British coal (supplied from Selby?), to £7.33 ($11) for the 120 millionth tonne (deliverable to Thamesside?)